THE HOMESTEADING HANDBOOK

THE ESSENTIAL BEGINNER'S HOMESTEAD PLANNING GUIDE FOR A SELF-SUFFICIENT LIFESTYLE

KELLY REED

© **Copyright 2021 - All rights reserved.**

The content contained within this book may not be reproduced, duplicated or transmitted without direct written permission from the author or the publisher.

Under no circumstances will any blame or legal responsibility be held against the publisher, or author, for any damages, reparation, or monetary loss due to the information contained within this book. Either directly or indirectly. You are responsible for your own choices, actions, and results.

Legal Notice:

This book is copyright protected. This book is only for personal use. You cannot amend, distribute, sell, use, quote or paraphrase any part, or the content within this book, without the consent of the author or publisher.

Disclaimer Notice:

Please note the information contained within this document is for educational and entertainment purposes only. All effort has been executed to present accurate, up to date, and reliable, complete information. No warranties of any kind are declared or implied. Readers acknowledge that the author is not engaging in the rendering of legal, financial, medical or professional advice. The content within this book has been derived from various sources. Please consult a licensed professional before attempting any techniques outlined in this book.

By reading this document, the reader agrees that under no circumstances is the author responsible for any losses, direct or indirect, which are incurred as a result of the use of the information contained within this document, including, but not limited to, — errors, omissions, or inaccuracies.

CONTENTS

Introduction   5

1. The Homesteading Spectrum   13
2. Is Homesteading Legal?   29
3. Homestead Utilities   47
4. Farming and Raising Animals   73
5. Medical Care   95
6. Homestead Maintenance   103
7. Homestead Budgeting   111
8. Different Types of Homestead Properties   125
9. Being Part of a Homesteading Community   141

Conclusion   151
References   157

A SPECIAL GIFT TO MY READERS

Included with your purchase of this book is your free copy of
Your Homestead Planner

Follow the link below to receive your free copy:
www.kellyreedauthor.com
Or by accessing the QR code:

You can also join our Facebook community
Homestead Living & Self Sufficiency,
or contact me directly via kelly@kellyreedauthor.com

INTRODUCTION

"When the world wears out and society fails to satisfy, there is always the garden."

— MINNIE AUMONIER

Do You Wish You Had a Better, More Fulfilling Life?

In our modern lives, it's so common to feel as though something is missing. Sometimes we struggle to keep up with our day-to-day existence, working frantically in draining jobs, feeling as though our lives are slipping by us, and having no time to actually enjoy anything. We may feel numb, as though something has sucked all the meaning out of life; even when we *do* have downtime, all we do is bounce from one screen to another trying to figure out what will give us a sense of purpose or a flicker of happiness. Whole bookshelves of self-help books have been published in an attempt to help us improve our lives, but what if they miss a crucial element of our dissatisfaction? How can we improve our lives from within a societal structure that prioritizes cheap consumption and doesn't care how detached we feel as long as the cogs keep turning?

We tend to beat ourselves up for not being able to better our lives, but what if we're not the problem? Ever-increasing numbers of people are realizing that they want something more fulfilling for their lives, and are seeking that fulfillment by changing the way they relate to the world around them. Seeking a greater connection to nature and a more intentional way of life, people are turning to homesteading. Keep reading to discover why.

What Is a Homestead?

What comes to mind when you hear the word "homestead"? Maybe it conjures up images of a lost past where our ancestors worked hard to eke out a living, staying much closer to the land, the animals, and the plants that live and grow on it than the people of today. Maybe you think of an idyllic sun-drenched farm where fluffy lambs frolic in green fields, and children learn independence by fetching eggs from the chicken coop and milking cows for the cream to make ice cream. Maybe you picture somewhere far from civilization, a lone building standing amidst acres of land. Maybe you see gardens in bloom and tables full of their harvest. Or, perhaps you see solar panels and irrigation systems, closed systems providing all the electricity and water you could need. None of these ideas or images are wrong—and yet, none of them tell the whole story.

A homestead is simply a place where people make a home for themselves—the keywords being "make" and "themselves"! Homesteading is about learning to meet your needs with things you grow, gather, or otherwise produce yourself. That might mean you have a vast farm in the heart of the country, or it might mean you have a small window garden in a tiny apartment in the city. The truth is, there are many ways of homesteading, and many types of people who homestead. If you've ever wanted to make your life a little bit more sustainable, more self-sufficient, more organic, or even just a little bit more independent, then homesteading has something to offer you.

Who Should Read This Book?

If you were curious enough to pick this book up, it's for you! Maybe you already know that you're interested and you're hoping to find a great book to guide you through the process. Or, maybe you came across this book and it sparked something in you, and you'd like to follow that spark and see where it leads. This book is meant to impart knowledge and distill a broad topic down to its most essential parts. However, it will also help you to imagine a new way of living and give you the confidence to pursue that new lifestyle. If anything on this list applies to you, you are one of the many people that I wrote this book for:

- You're a city-dweller who longs for a more rural lifestyle.

- You want to reduce your environmental impact.
- You want to be less dependent on others for your food, water, and power.
- You're tired of expensive power and water bills.
- You dislike that your food is highly processed and exposed to chemicals and pesticides.
- You wish you could spend more time connecting to nature, experiencing the benefits of outdoor life, and feeling confident and satisfied that you can take care of yourself.
- You want more control over your life in general.
- You're planning to start a life off the grid and you want to make the best choices possible.
- You can't go off-grid at this point in your life, but you'd like to make small changes toward a more self-sufficient lifestyle.

If you identified with any of those statements, then this book can help you on your way to building a better, more rewarding way of life.

Who Wrote This, and What Do They Know About Homesteading?

My name is Kelly Reed, and my husband Robert and I have been homesteading since 2009. Over a decade of experience with off-grid living has given us a wealth of knowledge and expertise, but it wasn't always like that! When we started, we were totally new to all of this, just like you. We spent a lot of time and energy looking through resources that promised to help us plan and prepare. We found that there were plenty of resources—in fact, there were almost too many—but it was difficult to navigate such an overwhelming amount of information. Sometimes, these sources seemed to contradict one another, and many also offered an idealized idea of what homesteading life was like, which set us up for a rude awakening. This confusion meant that we went through a lot of frustration and pain, not to mention losing a lot of time and money when we made uninformed or misguided choices. Through a lot of trial and error, we've learned the best way to do things—but we don't want rude awakenings or trial and error for you. That's why I've written this book. I've only included the things I believe you need to know so you won't have to sift through tons of irrelevant content. Reading this, you'll get the benefit of all of the years I spent testing my research and honing my knowledge, and I can't wait to share it with you.

Getting Started on Your Homestead Journey

There are many ways you could go about starting your homesteading life. You could jump right in, figuring things out as you go along. That may sound exciting and appealing; however, it could be costly if you make uninformed decisions or expensive mistakes. A large enough error might even lead you to regret your decision altogether. On the other hand, you could spend a long time planning by reading, researching, and consulting experts. However, after days, weeks, or months of wading through the enormous amount of information out there, you might feel no closer to having a solid plan. In fact, you might feel frustrated and stuck, and end up never actually getting your dreams off the ground. Neither of these situations is ideal.

A better option is to find a thorough, unbiased guide that will give you the right information and practical solutions—that is exactly why this book exists. We've done the hard part for you; this book compiles all the basics that you need to know to get started while dispelling common myths and giving you a *real* idea of what the lifestyle is like. It will empower you to make well-informed decisions and give you what you need to get started, whether you're hoping to start a garden in your apartment window box or build a self-sustaining farm from the ground up. Here are some of the things you can expect to learn in the coming chapters:

- what homesteading is
- all the different ways it can be beneficial and how to incorporate those things into your daily life, even if you don't live in the country or want to go off-grid
- what to expect on your homesteading journey
- how to save time and money by avoiding common pitfalls and mistakes
- the laws, regulations, and other requirements related to homesteading
- the best places in the United States for homesteaders and/or off-grid living
- how to operate your own utilities, including renewable water and energy systems
- how to prepare for medical emergencies and other disasters

- what to expect, and how to prepare for homestead maintenance
- how to budget, and especially how to homestead on a small budget
- the differences in the types of properties, and which is best for you
- the importance of community, and how to find (and be) good neighbors

Reading this book will not only give you tons of great information and practical tips—it will invigorate your imagination and open your eyes to the many possibilities of homesteading. And, best of all, it will give you the confidence and information you need to make the right choices for you. Now, let's get started!

1

THE HOMESTEADING SPECTRUM

Earlier, I asked you what it is that you picture when you hear the word "homestead." Hopefully, you're already beginning to expand your ideas of what that can be, and to understand that this way of life is a spectrum along which there is something for almost everyone, no matter their lifestyle or location.

Homesteading life exists on a spectrum.

However, some popular perceptions aren't actually true. Some people, for example, believe that homesteading means going without modern conveniences like electricity. A quick search online reveals a number of homesteading blogs, proving that many homesteaders have not only electricity but the internet as well. Below, we'll dispel some other common myths and misconceptions.

Myths and Misconceptions About Homesteading

Myth: Homesteading requires a large amount of land.

It's not the space that makes a homesteader—it's what you do with that space. You don't need a large amount of land or even *any* land. You can actually do this anywhere, and it can be scaled up or down to fit within the space available to you. Some people may choose to own acres of land where they grow crops and raise livestock. Others may choose to own smaller amounts of land or to start out by leasing land in order to gain skills and experience before becoming landowners themselves. And some may focus on other ways of being self-sufficient, such as making their own clothes and cooking all their own meals from scratch using foods bought locally. It's all about how you approach your life; there are many ways of increasing the sustainability of your lifestyle that have nothing to do with size.

Myth: Homesteading means you're living on free land.

This idea probably comes from a time in American history when the federal government had a lot of land that it had seized or purchased, and wanted to encourage Americans to settle on that land. To achieve this goal, several laws—collectively called the Homestead Acts—were passed in the late 1800s and early 1900s. These laws allowed people to move onto parcels of land, where they would live, farm, and otherwise care for the land. If they did this for five years, the government would grant them the deed to that land.

As appealing as the idea of free land in exchange for labor and care sounds, the last Homestead Act was passed over a century ago in 1916, and it isn't nearly as easy to find free land nowadays. There are still some places in the U.S. running programs that give people land for free or for extremely reduced prices, so if it's your dream to live on free land, don't despair—it can be possible to make that happen. However, no one should feel like they're not truly homesteading if they aren't living on free land.

Myth: Homesteading requires special expertise or prior farm experience.

While planning and knowledge are always good to have, you don't need any special skills to get started and you don't have to have a background in farming. In fact, historical homesteaders (which we discussed earlier) came from all different walks of life and learned as they went. Modern homesteading is no different; the most important thing you need is the passion to start a new way of life. That passion is what will drive you to learn and to try new challenges, which will help you grow and learn. Start small, work hard, and you'll soon find your knowledge and skills growing exponentially.

Myth: Homesteaders all live "off the grid," and homesteading is all about preparing for doomsday or some other apocalyptic future.

It's a common misconception that all homesteaders live "off the grid," meaning that they are not connected to any public utilities such as water and electricity. While many *do* live off the grid, this is a matter of opinion and preference. If you are able to live off the grid and that's what you want, then you'll find resources in this book to help you do that. But if you're not interested in a life off the grid or if it's not feasible for you at the moment, it is absolutely not a requirement.

This misconception goes hand in hand with another, which is that people who homestead and/or live off the grid do so because they are concerned about a future where "the grid" fails and we are all left to fend for ourselves. But while this lifestyle will certainly make you more independent and able to provide for yourself, most people don't seek it out because they're worried about future disasters. Instead, they're drawn to the do-it-yourself lifestyle, and enjoy the autonomy that comes from knowing how to build, cook, and otherwise create their homes and lives on their own terms.

Myth: Homesteading means you're completely self-sufficient.

Again, there are very few rules about this way of life, and as with other parts of it, the self-sufficiency that people practice exists along a spectrum. Just as some live totally off the grid, some may be absolutely self-sufficient. More common, however, are those people who value and appreciate community. It is rare for anyone to never need help from anyone else, whether that help is in the form of exchanging goods and services, giving and receiving advice, or sharing knowledge. Your community might be your local network or the broader community across the country, but either way, that community helps you thrive. It also allows people to pursue their own particular interests. One person might be an excellent gardener, another might be great at sewing, and a third be excellent at dealing with solar energy. Before our modern monetary system, people negotiated excesses and shortages through barter and trading, using other people's proficiencies to make up for their own shortcomings, and vice versa. Homesteading follows this philosophy.

Myth: Homesteaders live far away from other people, surrounded by dangerous wilderness.

As we've just discussed, there's a lot of community in homesteading. This includes many people who have neighbors within walking distance. And as will be discussed later in this book, there are many ways to reduce the risks of living in the country, whether it be receiving proper training, securing communication lines, knowing how to contact and access emergency services, maintaining your stock of living necessities, or having basic first aid knowledge. Safety is less dependent on your environment and situation than it is on being prepared for what that situation and environment might generate. For some, this approach to life brings an increased sense of safety and security because they feel more informed and prepared to take care of themselves in a variety of situations.

Myth: Homesteading isn't enough to support yourself or earn an income.

Actually, there are all sorts of ways that you can earn income in this lifestyle. Some people make money directly from their homestead, by selling either goods (things they've grown and/or made) or services (skills they've acquired and can share with others). Some work part-time or have seasonal jobs near where they live, and others work remotely via the internet. Some even share their life via blogs, which can be monetized. There is a myriad of possibilities, and because of their can-do spirit, homesteaders are imaginative and innovative, and they apply those traits to money-making. Plus, many of the costs associated with more conventional lifestyles are eliminated by a more sustainable way of living. By growing your own food, making your own clothes, and producing your own power, you can greatly reduce your cost of living.

Myth: Homesteading is very difficult and too time-consuming for you to have other jobs or hobbies.

To be completely honest, this life is unquestionably difficult at times. This book would not be giving you the unbiased and clear-eyed overview it promised if it claimed otherwise. Especially at the beginning of your journey, you may struggle to learn new skills, to adapt to a new way of life, or to balance incorporating these new ventures with the more modern elements of your life. But many, *many* people have come before you, and their choices illustrate the variety of ways to structure your life so that it works for you. Homesteaders have full-time jobs, they volunteer, they pursue hobbies, they raise families. At times, of course, there are setbacks, failures, and frustrations. Persevering through these things, and learning that you can, is a big part of what makes this life so rewarding.

Myth: Homesteading is very simple, and with minimal effort, you can create an idyllic paradise.

While some fear this undertaking is too difficult, others have an idealized view of it, picturing the enjoyment of the end product without considering the effort required to achieve those results. A homestead is not the garden of Eden—it takes work. This does not mean, however, that it is unattainable, or will require the sacrifice of your whole life. You will succeed by starting where you are and finding the right balance—and, in return, you will get closer to nature, become more aware of the world around you, know what goes into your food and clothes, and in general create a healthier, more holistic life for yourself. For those who truly love it, the effort is more than worth those rewards.

Myth: There is one true way to be a homesteader.

It should be evident by this point that there are exceptions to almost any "rule" that you might have come across. In fact, there is really only one constant: all homesteaders hope to become more self-sufficient. If you hope for that and you have the will to see that hope through, then you are in the right place. Homesteading is not about doing any one thing. It is about creating a life where you have the means and skills to meet your needs; it means regaining some of the self-sustaining skills that have been lost over generations and reclaiming a role in your own subsistence, rather than being merely a detached consumer; it means valuing not just the end product, but the process that created it; it is about building a more intentional life for yourself.

Different Types of Homesteading

While there is no one true way, most homesteads can be loosely grouped into one of several categories. These categories are based on two main factors: location and scale. In other words, where will you make your home, and how big will your operation be?

Apartment Homesteading

This is homesteading on the smallest scale, for those who live in urban environments, rent their homes, and/or own no outdoor space. Apartment homesteading is all about making the most of the space you have. It often involves a good mix of self-sufficiency and modern conveniences. Those who practice it focus on ways they can modify conventional apartment living to be more self-sustaining. If this sounds like you, here are some ways you can get started:

- **Look for ways to reuse items and materials.** This could be switching from disposable items (like paper towels) to ones that can be used many times (such as a cloth towel), or keeping items that can be used more than once (like glass bottles and jars). It could mean shopping at thrift stores rather than buying new clothes.
- **You could also try looking for tutorials that teach**

you how to repurpose items that you no longer use into something you need. Get creative—imagination is an important part of this!

- **Make your own things.** This can be as complex as learning to sew your own clothes, or as simple as nailing together some boards to make a basic shelf. You could learn to carve, to knit, to weave, to make your own cleaning and beauty projects, or all of the above.
- **Start a garden.** There are any number of ways to grow a garden, even if you have no outside space. You could have a container garden that sits on your windowsill, or your balcony, if you have one. You could see if there's garden space available on your building rooftop—if there's not, you could ask the landlord if you can start one. Some neighborhoods also have community gardens, where you will be given your own allotment where you can plant and cultivate vegetables, fruits, and herbs.
- **Cook and preserve your own food.** If you're used to eating a lot of restaurant food and takeout food, start by learning some basic cooking skills. If you already know the basics, try going deeper to learn to cook completely from scratch. And to make the food you're growing in your garden last longer, you can also learn how to can, pickle, dry, and otherwise preserve that food so it can feed you for months to come.

Backyard Homesteading

A step up from apartment homesteading, this takes place in an urban or suburban environment where you have your own yard that you can use to grow food and raise small livestock. It still usually involves a mixture of traditional living with the more conventional modern lifestyle. Backyard homesteaders make the most of their land by planning carefully, which allows them to grow lots of different plants in different seasons and for different purposes. This can be surprisingly effective and is usually limited only by your planning skills and your imagination (and occasionally, your homeowner's association—be sure to check the rules and incorporate them in your plans). If you're considering backyard homesteading, here are some ideas to get you started:

- **Raise livestock.** It might seem strange to think of chickens or goats in the suburbs, but given the right conditions, it's totally doable. Raising hens to provide fresh eggs is a great first step in raising livestock. If you have a family, feeding the flock and gathering the eggs are good ways to involve your children and help them learn independence.
- **Upgrade your garden.** Many homeowners plant trees, flowers, and shrubs in their yards, but that space can also be used to grow edible plants. You can plant nut bushes or fruit trees. Utilize planting techniques like companion planting or intercropping (both methods of growing different types of plants in the same space) to

make the most of the area that you have. Again, get creative!
- **Learn to fix things that break.** Owning a home means that you're responsible when things go wrong. Instead of calling a plumber, an electrician, or another type of handyman when something breaks, learn to do your own repairs. You can still consult professionals for big tasks, but by educating yourself about how to do small repairs, you can work your way up to more complex jobs and find more independence.

Preserve your own food, make your own things, and lean into reusable materials on a larger scale than you could in an apartment. More room means you can try things on a bigger scale. Use your basement to store canned and preserved goods. Hang your clothes to dry on a clothesline rather than using a dryer. Set up a basic carpentry shop in your garage or spare room. Thinking creatively is key to having a thriving suburban homestead.

Small- to Large-Scale Homesteading

This is probably the most well-known type of homesteading. These are often in rural locations and on larger areas of land, though the size can vary wildly. The smaller end of the spectrum might be a few acres while the larger end might be more than a hundred. The biggest difference created by scale will be exactly how self-sufficient you can be. If you want to raise large livestock like cattle, you'll need enough land to support them. If you want to grow their food yourself, rather than purchase it, then you'll need enough acreage to grow both livestock feed and whatever food you need. If your goal is to homestead on a large scale someday, these are some aspirations you might begin working toward:

- **Expand the type of livestock you raise.** If you have enough room, this might include raising cattle for meat or cows for milk or sheep for wool. It also might include building beehives to harvest your own honey or raising enough chickens that you will have a surplus of eggs to sell for profit.
- **Plant and harvest crops.** If your property is of a smaller size, you might have a large garden (or possibly several), some fruit trees or bushes, and potentially a greenhouse or two. If your farm is larger, you might plant things on a larger scale, with orchards of fruit trees and fields of crops. Again, at this point, you might no longer be growing food just for yourself, but also to sell for profit, or to exchange with others in return for their goods or services.
- **Build your own house.** Take a cue from early homesteaders and start with just a parcel of land. It's not necessary to build your own house to be a homesteader, of course, but for many people, being involved in the planning and construction of their home from the ground up is one more way to be intentional about how they live. Plus, being involved in the building of your house means that you understand

how it is put together and are that much more equipped to maintain and repair it.
- **Become completely self-sufficient.** If absolute self-sufficiency is your long-term goal, then this is the choice for you. With enough time, planning, and land, you can create a system that supports you completely, providing shelter, food, clothing, water, and power without dependency on any external institutions. Homesteads like this take a *lot* of work, including setting up and maintaining your own power grid and water filtration system, but the effort is worth it for those that long for this level of autonomy.

Looking at these three groups, is there one that speaks to you, one that inspires you? The point of presenting these categories is not to make you feel that you have to categorize your own efforts but to illustrate the range of options available. As you begin to plan, remember that your self-sufficiency doesn't have to be extreme or total. Start small and look for ways to modify your current lifestyle. Ask yourself: What can I grow? What can I make? What skills can I learn? As you start to answer those questions, the next section will help you create some initial goals.

Setting Homesteading Goals

Once you've gotten started and you're planning your first goals, ask yourself two questions: What is my eventual objective? And are these SMART goals?

When thinking about your eventual objective, try to picture the perfect life; what does it look like? You'll want to keep this picture in mind to be sure that the plans you make now will move you toward that objective. Knowing your end game will help you prioritize your goals and plan accordingly. It will also help you to not get distracted by what other people are doing. Staying focused on your goals and values despite what others are doing is key to constructing a life that will satisfy and support you.

Now, with these goals in mind, you can ask yourself if these are SMART goals. SMART is an acronym invented by consultant George T. Doran in the 1980s; it stands for Specific, Measurable, Attainable, Relevant, and Time-bound. These five principles help to make sure that goals are concrete and achievable rather than vague and overly ambitious. Here's a further breakdown of how each concept affects your goal-setting:

Specific: Specific goals name exactly what it is you hope to achieve. They are not ambiguous or conceptual. They help you focus on particular steps that, when taken together, will lead you to your goal. For example, instead of saying, "I will become more self-sufficient with my food," say "I will learn to grow vegetables in my garden that I can eat."

Measurable: Measurable goals have an endpoint after which you can definitively say that you have achieved that goal. This gives you the ability to know how close or far you are to achieving your goal and keeps you from abandoning goals because you're struggling to recognize how far you've progressed. Instead of saying, "I will learn to grow vegetables that I can eat," say "I will learn to grow tomatoes, beans, and kale."

Attainable: Attainable goals are goals that you can actually accomplish. This means that you have to consider your current skill levels and limitations when setting a goal. If you're an accomplished carpenter, making a table might be attainable, but if you've never picked up a hammer, you'll want to start with something smaller. If you've never gardened before and your goal is to grow tomatoes, beans, and kale, start by setting a goal of growing one of those vegetables.

Relevant: Relevant goals are related to both your lifestyle and your long-term plans. This is where it is really handy to know your eventual objectives. If you hope to someday work and live on a large-scale farm, then food goals that help you on your way to being completely self-sufficient are relevant. If you don't currently live on a farm and don't plan to in the future, then food goals that aim toward sustainability through other means (a local food co-op, perhaps, or a mix of farmer's market, home-grown, and store-bought food) might make more sense for you.

Time-bound: Setting a time limit for your goals not only gives you the motivation to achieve them but also makes it easier to measure your progress. If you know you want to achieve a goal within a certain period of time, you can break down everything you need to do and create an action plan for achieving that goal. For example, instead of saying, "I will learn to grow tomatoes, beans, and kale," say "I will learn to grow tomatoes, beans, and kale by the end of this year."

Next Steps

By now, you should have a good idea of what homesteading is and isn't, as well as the different ways that it can be done. With all that information, you can start to dream of your new life and what it will look like. In the rest of this book, you will learn more about how homesteading works and how it can work for you. Once you have a thorough understanding of what it requires and what it has to offer, you can use SMART goals to help you start to move toward that vision.

2

IS HOMESTEADING LEGAL?

If you've done your own research online, you might be frustrated by the lack of clear information about the legality of homesteading. As we've established, though some people might live off the grid, homesteading and off-grid living are not the same thing. However, online sources often confuse the terms or use them interchangeably. This only adds to the difficulty in understanding what is and isn't legal. Further complicating matters is the fact that each state has its own laws and regulations regarding both homesteading and off-grid living. But don't be discouraged! This book exists to inform and guide your decisions, and that includes helping you understand what is and isn't legal. If you plan to buy land, build a house, set up your own water and electricity, raise livestock, or grow your own food, understanding the laws discussed below will be especially important. We'll walk you through them one by one.

Camping on Your Property

It may seem strange since you own the land on which you would be camping, but in a lot of places, there is a limit to how long you can camp—usually two weeks. After that two weeks is up, you must either move into a more permanent abode or move off the property. This includes camping in camper vans and some trailers. If you were hoping to camp on the property while you built your house, you can apply for a special permit to do so, but be aware that it can be very difficult to procure one of these permits. As with everything on this list, it is wise to double-check the exact rules of your local municipality, but know that it is unlikely you can just pitch a tent and call it a day. Eventually, you will need a permanent shelter to stay within the law.

Minimum Square Footage Laws for Houses

There are quite a few rules that govern building a house, no matter where you build it. Anytime someone builds a new structure, the plans for that structure must be approved by the local government. To start, there is a minimum required square footage. In other words, if your planned house is too small, you won't be granted building permits. The minimum frequently falls between 500 and 1,000 square feet, meaning that many micro-houses or "tiny houses" might not be approved.

Minimum Square Footage Laws for Land

In addition to regulations on the size of houses themselves, there is also a minimum on the size of land you can buy (or sell). This minimum varies by location and environment, meaning that lots in the city are often allowed to be smaller than lots in the country. Local governments put these regulations in place to keep rural areas from becoming too densely populated. It is common for lot sizes in rural areas to have a minimum of five to ten acres, although in some places the minimum can be as high as 20 or even 40 acres. For both minimum house and land square footage, exceptions called "variances" can be granted to allow for smaller structures or lots, but these exceptions are very rare, and it is best not to plan on getting one.

Building Codes

Once you start building your house, you must follow national and international building codes. It is important to follow these codes because your building will be inspected by the local authorities to be sure it complies. It is also important because these codes are designed to ensure that structures are safe. If you're intimidated by the idea of keeping track of construction and being sure you are following code, you can hire a building contractor, whose job involves making sure you adhere to the codes.

Water Utility Regulations

The first rule with water is that you *must* have it! If your building plans don't include access to clean water, they will not be approved. Normally, this would be done by connecting to city- or county-run water lines; in some places, connecting to this network is required. Those wishing to live off the grid will look for locations where they are allowed to use alternatives to the municipal water supply, such as a well dug on the property or use of a natural spring. Another alternative is to collect and purify rainwater, but this method of getting water is frequently illegal even in locations where other alternatives are allowed.

Power Utility Regulations

Laws in most places do not allow people to live completely without electricity. This is generally because electricity is required to meet certain safety standards. As with water, some places will require citizens to be hooked up to the local electrical network. As a general rule, the closer you are to a town or city, the more likely it is that you will be required to connect to the power grid. In places where you are allowed to generate your own electricity, you can consider different options, such as solar panels, wind turbines, or thermal wells. If you do generate your own electricity, you might be allowed to sell any excess electricity you generate to the county or city, which it will then sell to other residents. Some states also incentivize the use of renewable energy with tax cuts. If you do opt for an alternative method of power generation, it is also important to be sure that it is installed and maintained correctly.

Septic System Regulations

One aspect of living "on the grid" that is often taken for granted is septic waste disposal. It is obviously illegal to dump septic waste, so if you're not planning to be hooked up to the local sewer system, you'll need to establish your own septic system. This involves testing the ground's absorption rate, getting the appropriate permits, and procuring and burying a septic tank.

Often, you will be required to have the system installed by a professional. There are some alternatives to using a traditional septic system, such as using a chemical toilet (like those used in RVs) or a composting toilet. However, these are only legal in some situations, and will also need to be approved before being put to use.

Raising Livestock and Growing Food

If you are planning to keep animals on your land, you must be sure that your property is zoned for agriculture. There are generally fewer regulations on growing food, especially fruits and vegetables. However, for both livestock and crops, if you intend to sell any of what you raise or grow, then you must procure a permit. Your land and any relevant facilities will be inspected before the permit is issued. With grain, in particular, it is important to determine if your operation will be considered a commercial farm, as it might then be subject to certain farming regulations. A related issue is game hunting, which also requires a permit (even if you're hunting on your own land) and the type of animal you can hunt may be restricted by location or season.

Paying Taxes

While many people choose to homestead because they want more control over their lives, there is one area where you will always be subject to local and federal jurisdiction, and that is taxes. No matter how independent you are, if you own property, you must pay property taxes, and if you make money, you must pay income taxes. It is always possible that you will qualify for certain tax incentives and exemptions, however, so it may be wise to consult a tax expert during your planning.

Choosing Where to Homestead

You may have noticed a common theme in all of the above information—the laws change depending on where you live. It is always important to double-check local rules and regulations, but if you're planning to relocate, then you have a chance to pick which state is the best place for you. The next section of this chapter is dedicated to helping you determine that.

A Brief Overview of U.S. Climate Zones

In addition to local laws, another very important factor to consider when relocating is what the weather will be like in your future home. Climate is one of the most important things to consider when planning for any lifestyle, but it is especially important when you plan to be working outdoors, caring for livestock and plants, and generally more exposed to weather and elements of nature. Specific needs might also be highly influenced by climate, such as how much sun you can expect to get if you plan to rely on solar energy. There is no one correct climate, as people can thrive in all sorts of environments, and some people are particularly suited to even the most extreme climates. However, many people find that mild, variable climates are ideal for homesteads, given their diverse seasons and lack of severe temperatures.

The continental U.S. is typically divided into these five climate regions:

The Northeast: Known for its seasons, especially its epic fall foliage, the Northeast encompasses the states of Maine, Vermont, New Hampshire, Massachusetts, Rhode Island, Connecticut, New York, New Jersey, and Pennsylvania. The climate here is varied and diverse, with chilly winters and semi-humid summers. Generally, the farther north you travel, the colder the winters; the farther south, the more humid the summers.

The Southeast: Containing the greatest number of states, the Southeast includes Kentucky, Tennessee, Virginia, West Virginia, Delaware, Maryland, North Carolina, South Carolina, Mississippi, Alabama, Georgia, Florida, Louisiana, and Arkansas. The Southeast is considered subtropical and can be quite humid. The winters are usually mild and the summers can be very hot.

The Midwest: Spanning the middle of the country, the Midwest is made up of Ohio, Indiana, Illinois, Michigan, Minnesota, Wisconsin, Iowa, Missouri, North Dakota, South Dakota, Nebraska, and Kansas. The summers here are often similar to those in the Southeast, hot and humid, but the winters get much colder. Snow and subzero temperatures are common.

The Southwest: Containing only four states—Oklahoma, Texas, Arizona, and New Mexico—the Southwest still spans a good amount of land. This area of the country contains a lot of deserts, which means the temperatures tend to be extreme. The climate in coastal areas, influenced by the currents from the Pacific Ocean, has more moderate temperatures that stay consistent throughout the year.

The West: This region includes states along the Pacific coast and in the northwest corner of the country, and is composed of California, Nevada, Washington, Oregon, Idaho, Montana, Wyoming, Colorado, and Utah. Summers here are often dry and cool, and winters also tend to be mild. The further south you go, the dryer the climate gets; likewise, the further north you travel, the better the weather becomes.

Hawaii & Alaska: Because they are not physically connected to the rest of the U.S., both these states have distinct climates from any other regions. Hawaii has a tropical climate, though local weather can vary quite a bit depending on altitude and land features. Alaska, because of its large landmass, is made up of many different climate zones. Most of Alaska's population lives in the southern and eastern parts of the state, which are milder in climate compared to the further, northern reaches of the state.

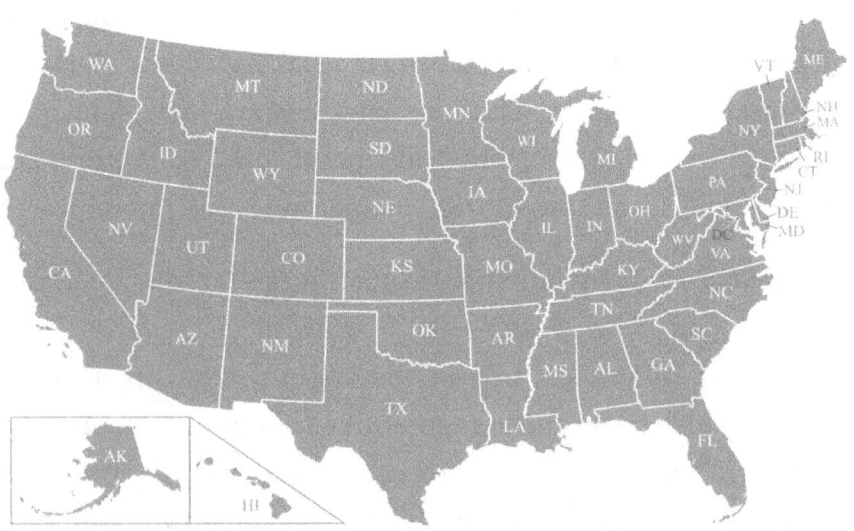

The Best U.S. States for Homesteading

As with all the choices you'll make in this endeavor, it is important to remember that the best state for your homestead is the one where you can build the life you want—even if that's not the place that the majority of others choose. However, several states jump to the top of consideration because of their suitability for homesteading and living off the grid. To make things simple, we've done the legwork for you, compiling and cross-referencing research into the handy list below.

Some key things to be aware of as you consider these options are land prices, cost of living, local laws and building codes, taxes (property, income, and sales), and, as we just discussed, climate. If your budget allows, it can be a good idea to visit any prospective states, specifically the area where you might like to live. Look around, talk to locals, and get a feel for what living there might be like. A state may appeal to you on paper, but leave you underwhelmed in reality. Alternatively, a state may have stricter laws, harsher weather, or other potential negatives, but you might decide these things are worth it if you really love the idea of living there. The key is to know both what you want and what a state has to offer, so you can choose a location that is the best fit for you.

Top Choices

Missouri

Missouri comes up frequently when researching homestead-friendly states. This midwestern state has varied landscapes and allows you to experience all four seasons without also experiencing the extremes of some other places. It has a mild winter that doesn't last too long, with the last frost happening in April or early May. This works well for growing crops, as does the average of 40 inches of rainfall per year. Missouri authorities are supportive of off-grid living, and their laws reflect that: you are not required to connect to a septic system, you're allowed to collect rainwater, and there are not many strict zoning regulations or building codes. Fishing is legal in many of Missouri's lakes, as well.

Missouri has a long history of farming, and most of the nearly 95,000 farms there are smaller and family-owned. This translates into quite a few communities focused on self-sufficiency, as well as a high amount of things like farmer's markets or "U-Pick" farms. You can declare your home a homestead after living on it for 40 consecutive months, and once declared, you can claim up to a $15,000 exemption on it. Other positives for the state include moderate taxes (both sales and property), incentives for using solar power, and an abundance of rural countryside. However, it is important to note that Missouri has high state income taxes, and land prices can be expensive. It is also good to be aware that droughts and other natural disasters do sometimes occur.

Oregon

This state in the far northwest of the country is known for its gorgeous scenery and prolific farmer's markets. Oregon is a great place to move if you like the idea of a lot of community, either through lots of close neighbors or as part of a town dedicated to homesteading. In fact, some communities in

Oregon have taken this to a whole new level, with large off-grid communes like Three Rivers and Breitenbush Hot Springs even becoming tourist attractions. There is a lot of diversity in Oregon landscapes, with deserts, forests, and beaches in different parts of the state. Farmland is usually inexpensive and, especially in the rainier parts of the state, very fertile.

Oregon offers homestead exemptions of up to $40,000 for single people and $50,000 for married couples. Within the city limits, you are allowed to protect up to one city block, while outside those limits, you can cover up to 160 acres. There are a lot of natural resources, such as high-quality timber, as well as a lot of public land for recreation, with very few restrictions on hunting and fishing. Other benefits of living in Oregon include no state sales tax, average property taxes, rebates for solar power, and programs for farmers like grants and educational opportunities. However, Oregon does have a high cost of living and fairly high income taxes.

Tennessee

Tennessee is a great option if you're on a budget and hoping to keep costs low. They have a lot to offer financially, including a cost of living that's 10% below the U.S. average and very low property taxes. The laws and regulations in this state are also appealing, with minimal building codes in a lot of counties and low restrictions on activities like rainwater collecting and raw milk herd-sharing. Tennessee also offers a rural homesteading land grant and hosts the annual Great Appalachian Homesteading Conference. The state protects property from being seized by creditors in the case of a financial crisis and allows exemptions of up to $5,000.

Tennessee also has a lot of natural beauty and great natural resources, including many opportunities to hunt and fish. The weather is temperate and the growing season lasts around nine months. The land is great for growing crops, especially in the middle and western parts of the state. Tennessee's main drawbacks are weather-related, given the occasionally cold winters and the possibility of tornados and flooding.

Idaho

Idaho is great for agriculture, having a reputation for some of the most fertile soil in the U.S. This explains why around 15% of the population are farmers. It also boasts beautiful rolling green hills and mountains. Its population density offers a good balance, as the population is sparse enough to allow you space and privacy without feeling truly isolated from other people. Both living costs and land costs are low, and natural disasters are rare. The laws in Idaho offer support for growing crops, raising livestock, and even homeschooling your children, as well as giving declared homesteaders who own their property financial protection up to $100,000, and exemptions in property taxes. This explains why there are over 60,000 homesteads in the state.

Idaho's other advantages include outdoor recreational possibilities (such as skiing, whitewater rafting, fishing, and hunting), a low crime rate, and tax incentives for solar power. The state is becoming more popular, though, so there is potential for a population boom that would drive up prices; other potential negatives include harsh winters, high income taxes, and occasional floods, wildfires, and earthquakes.

North Carolina

North Carolina might be less popular for those who want to farm lots of crops, but it offers great opportunities for other types of homesteading. Its mountains are well-suited to raising goats and there are lots of wild foods, such as berries or mushrooms, for those who are interested in foraging for food. The western part of the state is particularly popular among homesteaders, with major communities in the Saluda and Black Mountain regions that continue to grow.

North Carolina offers solar power rebates, low property taxes and laws that are conducive to homestead and off-grid lifestyles. Its growing season is a good length and the climate is moderate. The state also gives declared homesteaders good protection under the law, which might offset some of the negatives, such as high sales and income taxes and high land prices.

Wyoming

Unlike Oregon and North Carolina, which offer a lot of community, or Idaho, which offers a good balance between privacy and community, Wyoming is a great state for those who really value solitude. The broad horizon and extensive space mean great views and, combined with the low population density, a chance to withdraw more completely from modern life.

Wyoming has a lot of resources for both farming and ranching, and the land—good for both growing crops and raising livestock—isn't too expensive. With a low cost of living, low property and sales taxes, and no state income tax, Wyoming is also appealing to those on a budget. Additionally, exemptions for homesteaders protect up to $40,000. The state boasts an average of 114 sunny days per year, which is higher than states known for their sunshine, like California. Wyoming is not for everyone, however, because of the severe winters, short growing season, and lack of rain (which can occasionally lead to wildfires).

Other States to Consider

The above states may be the most appealing to homesteaders for their laws, climate, and land, but they are not the only options. Read on to discover other states that still have a lot to offer.

Alabama — Alabama's cost of living is low and the cost of land also tends to be low. It has some of the lowest property taxes in the U.S. and therefore is ideal for large land purchases. The state gets 56 inches of rain per year and there are no restrictions on collecting rainwater. Some places in Alabama have no building codes at the county level. However, Alabama can be quite hot and occasionally has hurricanes or tornadoes.

Alaska — Alaska is similar to Wyoming in that it will appeal to those who wish to have a lot of privacy and solitude. In fact, some parts of Alaska can only be reached by plane. However, there are parts of Alaska, especially further south, that have more established communities. Alaska also has a low level of regulation and generally permissive laws. The trade-off for this is an extreme climate, especially in the north of the state, where some regions are inside the Arctic Circle.

Arizona — This state in the Southwest is warm all year and gets the most sunshine of all U.S. states, which makes it ideal for those who want to use solar power. The land is also very cheap and the growing season is long. However, much of the land is desert which can make growing food a challenge, especially as Arizona can have droughts. Locations near fresh water are ideal, as this allows you to drill for a well and get the water you need from there.

Arkansas — Arkansas has a lot of positives including an unlimited homestead exemption, low to moderate land prices, low property taxes, and a low cost of living. Known for the grandeur of the Ozarks, the state also has 9,700 miles of streams and 600,000 acres of lakes. With such expansive wilderness, this is another state that's great for those who want a little more seclusion and those who intend to rely on natural resources. The climate is mild and the growing season is long, lasting 200 days. It can be harder to find land here, though, because of competition with bigger agriculture corporations, and the income and sales taxes are high.

Colorado — Flexible zoning codes and local municipal support of green and sustainable housing are two of the big draws in Colorado. Some counties have no building codes, the most well-known of these being Delta, Custer, and Montezuma counties (but note that the state itself still has building codes that must be followed). Colorado is also a great place for renewable energy sources, with plenty of wind and sunshine for both solar and wind-powered electricity.

Hawaii — Hawaii's weather and tropical climate have made it a famously nice place to live, but lesser-known is the fact that outside of cities, many Hawaiian residents already live off the grid. The state gets a lot of both sunshine and rain, which means its soil is great for growing things. Wind and solar energy are both good options, and there is no restriction on rainwater collection. The state does have some restrictive regulations, however, as well as a high cost of living.

Indiana — The southern portion of Indiana is especially popular among homesteaders as it has a low population density, a good growing season, and warmer weather. Land can be expensive in Indiana, but taxes and living costs are generally low. There is a tradition of farming in Indiana with the majority of its 56,000 farms being smaller scale and owned by families. Indiana also has a lot of green space; unfortunately, it also has frequent floods and tornadoes.

Iowa — The combination of great land and lower living costs makes Iowa a favorite among farmers. Low property taxes and a 100% exemption on homesteads up to 40 acres in rural areas also make it appealing. If you choose to install solar panels, the cost of that is exempt from sales tax, and you're also eligible for a tax credit. It is worth being aware that Iowa, like some other states, can have higher farmland costs because of its popularity with farmers. It also has higher state taxes and residents sometimes have to deal with tornados, floods, and long winters.

Maine — Maine has many homestead-friendly laws and zoning regulations as well as many natural resources such as timber, water, and rock. There is a lot of open land and prices are good, especially in more rural regions. Many people admire Maine's landscapes and the variety between the inland and coastal areas. The growing season is unfortunately short, but other benefits, such as low property taxes, still draw homesteaders to the state.

Michigan — While Michigan is a relatively expensive state to live in and has stricter regulations than many other states, it also has quite a lot to offer. It has good soil and right-to-farm laws, meaning farmers are protected against nuisance complaints from neighbors so long as they follow approved farming practices. This might be why Michigan has over 47,000 farms of varying sizes. The state also has Lake Michigan, in which residents are allowed to fish for trout and salmon. The growing season is around 140 days long, though it is common practice to extend the growing period by building greenhouses.

Montana — With its wide, sweeping plains and prairies, Montana is great for livestock and wind energy. Outside of those grasslands, you can find timber and other natural resources. The land is affordable and the cost of living is decent. Montana's growing season is short, though, making it another state where people tend to build greenhouses. With a sparse population, Montana is another option for those looking for space and privacy.

Ohio — Ohio is very amenable to off-grid lifestyles, with many counties having such relaxed laws and zoning codes that they don't even have a dedicated permit office. Other points in Ohio's favor include inexpensive land costs, low property taxes, low cost of living, lots of natural resources, and a five-month growing period.

Texas — Most parts of Texas have a long growing season and the land can be very reasonably priced, especially in less populated areas. Additionally, Texas offers a lot of timber and rock, which residents can use for construction. Something to be aware of, however, is that Texas contains many desert regions, which means a scarcity of water sources. This can make farming more difficult and cuts down on opportunities to hunt or fish.

Virginia — While Virginia's high population may make it a poor fit for those seeking solitude, the state has many positives if you're looking for a less remote lifestyle. It has good soil quality and a very long growing season. Unfortunately, cost of living, income tax, and land prices tend to be high, but property taxes are low and the climate is mild. The higher population also means a higher density of education and work opportunities if you're interested in a way of life that balances community with independence.

Next!

As you can see, there's a lot to think about if you're considering moving for your homestead! Once you've found the best place for it, the next two chapters have everything you need to get started setting up your utilities, planning a garden, and raising livestock.

3

HOMESTEAD UTILITIES

This chapter will walk you through some of the most important aspects of setting up your homestead: how you will take care of your water, power, and waste needs. All three are crucial systems not only because they are everyday necessities, but as we established in the last chapter, most homes are legally required to have access to water, functioning electricity, and an appropriate place to dispose of sewage.

Water Systems

Many factors will affect the type of water system that you choose to use. You must consider the type of property you own, the average rainfall in your area, and the natural water sources available (or not) on your land. If you're buying a property, it will be important to find out what sort of water resources exist on the property through the seller/real estate agent as well as your own testing and research.

Your household's average water consumption and budget will also influence your choices. Keep in mind that having a garden or livestock will increase your water consumption; if your budget is tight, learning to reduce water usage where you can is very helpful in keeping water costs low. It's a good idea to have multiple water supplies in the event that one fails, especially if living off the grid. Below are some of the most common water systems that homesteaders use:

City Water ("on the grid")

This one is pretty self-explanatory. If you're not interested in living off the grid, you can connect your home to the local municipal water supply. This method has the least amount of control, but also the least amount of responsibility. If you're living in a town or city, you may even be required to use this water supply, but this requirement becomes less likely as your location gets more remote.

Well Water

Most people are familiar with the basic function of a well. It is essentially a hole dug down from the surface that allows you to access underground water. Wells are expensive to establish, but once set up, they don't require much maintenance and they're one of the most reliable ways to get water. If you're buying a rural property that has already been inhabited, you may be lucky enough to find a place that already has a well. This can make your life easier, but always be sure to have the well water tested before you buy. The Environmental Protection Agency has a list of recommended substances to test for in water, including bacteria, nitrates, pH levels, dissolved solids, tannins, chloride, and copper, as well as water hardness. This testing should be done by a lab that is certified at the state level; you can usually get information on which labs to use from your local health department. (You should also continue to test your water every year or two to monitor its levels.)

WATER WELL CONSTRUCTION

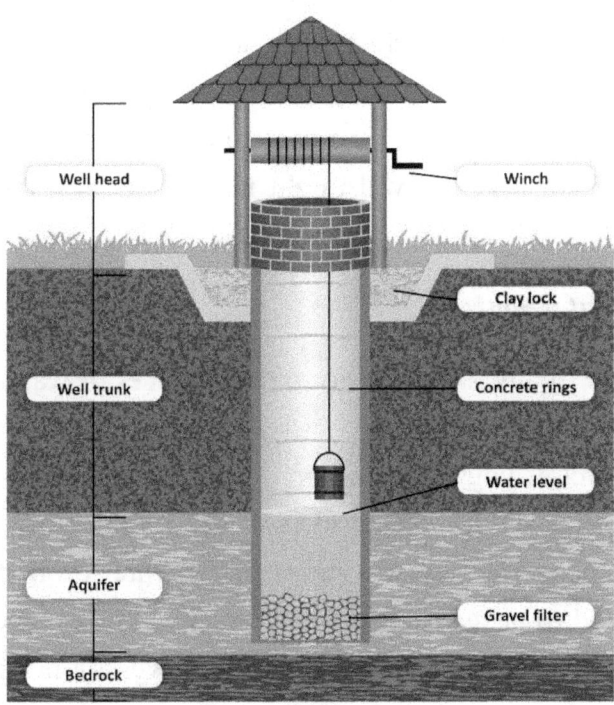

If you don't already have a well on your property, you'll have to drill one. Wells come in a variety of depths; shallower wells are cheaper and easier to draw from, while deeper wells have less risk of contamination, but are more expensive. Once the well is dug, a pump draws the water up from the ground. (You can draw the water up with a dipper or bucket, but that method is very slow because of the limited amount of water that can be transported at one time.) It is standard to use electrical pumps nowadays, although some people do prefer a manual pump, or at least have one set up in case of a power failure.

Before digging a well, check to be sure you'll have the rights to any water you find. Even if it's on your property, if the water connects to other properties in the area, you may not have full irrigation rights. Lastly, if your area is known for seismic activity or other activity that destabilizes the ground (such as drilling or fracking), a well might not be the best choice. But there are plenty of other options to consider:

Natural Water Sources

This refers to any naturally occurring sources of water, such as a pond, creek, or river. Obviously, you can only use a natural water source if you have one on your land (and even then, be sure you check with the local laws—sometimes owning land does not mean you own the rights to the water on that land). It is also good to be aware that many creeks or streams dry up in the summer. Until you have spent a few years on a property, don't trust the claim that any body of water is year-round.

Each of these water supplies has its own advantages and drawbacks. A creek or river is not only a water source but could also be used for a hydro-power system (more on this later). Running water also tends to be purer than still water because the movement through rocks and sand acts as a filter system. A pond, however, can be duplicated with DIY efforts. Some homesteads even use a man-made reservoir to collect rainwater runoff. Springs, which are essentially naturally occurring wells, are great because they bring water up from deep underground, which can then be diverted into your home's system.

Rainwater Collection

Obviously, rainwater collection works best in climates that see a lot of rainfall. As noted above, many people will use man-made reservoirs to collect rainwater, usually by redirecting runoff from their roof. Rainwater is free and it can be easy and fairly inexpensive to set up a system to collect it. However, because this supply is dependent upon the weather, it can be advisable not to make this your primary water access. This will depend on how much rain you can expect to get and your capacity to store excess rain for use on dry days. And, as noted in the previous chapter, sometimes rainwater collection is illegal, so be sure to check that with your local laws.

If you do plan to use your roof runoff, it's important to know that the material of your roof makes a big difference. Water that has been collected from an asphalt roof, for example, can be *very* contaminated, suited to watering plants but not drinking. Tile and slate roofs are considered much better, and metal roofs are considered ideal for collecting rainwater runoff.

Water Storage and Delivery

In addition to collecting water via whichever methods are available to and fitting for you, you'll need a way to store any excess, as well as a way to deliver that water to all the places that you need it. For the first, people commonly use cisterns, and for the second, you'll most likely need a pump.

Cisterns

A cistern is merely a tank specifically for water. These can be either above or below ground and are usually made of plastic, which is inexpensive, light when empty, and resistant to microbial growth. Above-ground cisterns tend to be smaller and easier to move, and sometimes double as containers for water transportation. Those below ground are often made of more durable material and are more expensive. If you live in a colder climate, it is best to have an underground cistern placed below the frost line, so that it will not freeze. If you have the opportunity, you can place a cistern so that your water pressure is aided by gravity.

Water Pumps

As just mentioned, some systems rely on gravity for their water pressure. This is called a "gravity-fed" system, and it depends on placing your water tank higher than your home so that gravity pulls the water down into your pipes. (This same principle is used in on-grid water sources, such as water towers.) This is only possible in some situations, and you may still need a pump to get water from your source into your cistern. Your other option is to rely primarily on a water pump, which can be stronger and more consistent but will require power. How much power will depend on your water consumption, as that will determine the amount of water you need to move on average. Some homesteaders use solar power for their water pumps, and we'll discuss solar power in more depth in the next section on electricity.

Water Purification and Filtration

Another thing that you need no matter which water source you choose (unless you're staying connected to the grid) is a way to filter and purify your water. You may see these terms used interchangeably, but filtration is used to mean getting rid of physical contaminants while purification is used to mean the removal of chemical contaminants, such as bacteria and other biological hazards. There may be times when you'll need to use very stringent filtration and purification methods for your water, such as water for bathing and drinking. There are other times when less exacting methods, or even no methods, will be needed, such as for watering your garden. There are laws around the purity of drinking water and you're responsible for testing your water at least twice a year (in the spring and fall) to make sure it is safe.

There are two basic types of filtration systems: inline and gravity-fed. Again, gravity-fed systems rely on the force of gravity to move water: the water is placed in the top of a container with a filter at the bottom, and gravity pulls it through that filter to another, clean container underneath. Gravity-fed filters can be used even without regular plumbing. Inline filters, on the other hand, are placed in your plumbing, meaning that the water is filtered as it enters your home.

It can be wise to employ multiple filtration and purification systems, especially if you have reason to believe your water sources might be contaminated. (Surface water, for instance, is particularly vulnerable to contamination, and can be unsafe to drink even if it tests okay.) Having one system that filters out particles and other physical contaminants and a second to remove biological hazards can be the safest way to go. It is also a good idea to have a backup method of water purification in the event that your main system is offline. Below are some of the different ways you can filter and purify your water.

Filters

You can buy filters and you can also make your own. Simple DIY filters, such as creating a membrane filter by straining water through a piece of cloth, are great to have on hand for emergency use, but are obviously too small in scale to filter all the water for your household needs. More involved processes include pumping your water through a carbon (bio) or ceramic filter.

Biofilters

In a biofilter, water moves through three layers to filter your water.

Biofilters are slightly more complex filters that can still be made at home. They involve three layers of filtration: gravel, sand, and charcoal. The gravel and sand work to screen out physical contaminants while the charcoal gets rid of chemical contaminants. (However, biofilters do not remove heavy metal or bacteria from your water, so be sure to keep that in mind when planning.) Gravel and sand are obviously quite easy to procure, and the charcoal can be made from waste wood gathered from your property. Make sure you clean these materials before putting them into your filter by rinsing them in a bucket of water. Swirl the water around until it gets cloudy, then drain it out and replace it with fresh water. Repeat until the water stays clear. In addition to cleaning all the elements before you put the filter together, this type of filter should be taken apart to clean the vessel and replace the filtering components yearly.

Ceramic Filters

Ceramic filters use the porous nature of clay to screen out impurities. You can still put these together yourself, although unless you are especially good at pottery, you'll need to buy the ceramic filters. If you're particular about the taste of your water, or you'd like an extra level of purification, you can get ceramic filters with cores made from activated charcoal or silver. This method, therefore, has more upfront costs (you need about two filters per person in your household just for drinking water purposes), but the filters can last up to a decade. Another potential drawback is that water filters very slowly through these systems, so it can take a while for water to be available after filters have been replaced.

Chemical Purification

This method adds chemicals to your water to destroy impurities. Generally, chlorine-based bleach is used. (Usually, when people talk about chlorine, they are referring to bleach with a chlorine base, though there are other bleaches with different chemical makeups.) As you might guess, one of the bigger downsides of this is the taste it gives your water. The power of this method of disinfection is determined by variables such as the temperature, pH, and clarity of the water. This is an easy and fast way to purify water, especially in an emergency. Some people also purify with iodine, but given iodine's light sensitivity, the potential for iodine allergies, and the fact that it is not safe for pregnant women, I don't personally recommend it.

UV (Sunlight) Purification

UV light bulbs emit frequencies that destroy contaminants in water.

You can purify water using actual sunlight or specially-made UV light bulbs. The frequencies in UV light can destroy the cell structures of contaminants in your water. This method of purification does not remove sediment or other debris from your water, and it is only effective against certain chemical contaminants, so it is best to pair this method with another process. It's also very important that there is nothing between the UV light and the water—even things that look clear to us can block UV rays. It is best for water to be uncovered so the light can hit it directly, but if you must cover your water, you'll need to do so with specially-made plastic that is designed to let UV rays pass through it.

There are two commonly used types of UV light bulbs: low pressure/high output (LPHO) and medium pressure/high output (MPHO). LPHOs are more energy-efficient, but less powerful and therefore require more units or more time to disinfect the same amount of water. MPHOs, on the other hand, are more powerful but less energy efficient. There are also low pressure/low output (LPLO) bulbs, but they are not as frequently used because they are much less powerful. Essentially, you'll always be trading off power for energy efficiency, and which you choose should be based on the volume of water you expect to disinfect and whether or not speed and power are more important to you than saving energy.

Distillation

Mimicking the natural water cycle, distillation works by heating water until it evaporates and then condensing the steam into water in a new, clean container. The heat of this process will kill bacteria and other microorganisms, while the evaporation leaves behind physical impurities like sediment and particles. Because of this, distillation is a very effective way of disinfecting your water. Technically, you can use distillation anywhere that you have a heat source and a way to collect the condensing water, including your kitchen. The smallest water distillers are actually made to sit on your counter or table, and they distill just a small amount of water at a time.

However, in order to be practical as a method for purifying all your water, you'll need a bigger operation. You can either have a distillation point installed in your plumbing system, or you can purchase a large (usually called "commercial") distiller. The in-system distiller purifies water as you need it, which means you are never distilling more water than you need. However, it does have an upfront cost of up to $1,000 for purchase and installation. A commercial distiller will have the capacity to distill around 75 gallons of water each day, a much higher volume of water than a countertop distiller, but it will use a proportionately higher amount of energy. Some people use a solar water distiller in order to distill a larger amount of water while using less energy. This involves placing water under glass or plastic with access to sunlight, which will then heat the water to evaporation. This method of distillation obviously is best suited if you live somewhere warm with strong, direct sunlight.

Boiling

As noted above, heating water will kill bacteria and microorganisms, so boiling your water will disinfect it. The water must be kept at a rolling boil for at least a minute. Notably, you'll want to pair a method of filtration with boiling, as on its own it will not remove physical contaminants. While not practical as a main source of purification, boiling is very handy for purifying small amounts of water quickly, or if you find your other methods of purification unavailable. It is also easy to do and, as long as you have a pan and a source of heat such as your kitchen stove, it's free.

Now that we've covered the most common water systems for homesteaders, let's take a look at the options for electricity.

Power Systems

When you're accustomed to living on the power grid, you might take power for granted because it's always on and available. But if you're going to generate your own power, you suddenly have to consider all the things in your home that require electricity, from the lights to the Wi-Fi router to your hairdryer. The first step when moving to your own power system is to consider how you might slim down your electricity usage. Just like with water, the less electricity you consume, the more your budget will thank you. Look for small tasks that you can do manually, and if you're building a new home, make sure to consider ways the design and construction can reduce your energy footprint.

Solar Energy

If you live in a climate that gets a lot of sun, solar energy can be a great option. Solar energy is very efficient, and once installed, it requires only a small amount of maintenance. However, you'll need solar panels to collect the sun's energy, an inverter that converts that energy, and batteries to store it. The price of this system and the cost to have it professionally installed is high, so you'll want to be sure you'll get enough sunny days to pay off this investment. Solar energy is also an option for heating your water, which is important if you don't want to take exclusively cold showers. (We'll discuss more heating options a little later in this chapter.)

Wind Energy

This is another source of power that depends on your climate. You should be able to look up the wind resource map for your area to find out what sort of wind your homestead will get. Wind energy isn't quite as efficient as solar energy, but if you live somewhere that is windy and overcast, it can work much better than a solar panel. The main costs for wind energy are the tower and turbine. It's important to have these somewhere relatively unobstructed; the higher the better, so if you have hills on your property, you might consider setting them up there. Because the towers are quite tall, you may struggle to be approved for them, especially if you live in less rural areas. Something to consider with wind energy is that the energy is generated by the mechanical process of turning the turbine, so the components will wear down over time and eventually need to be replaced.

If you're located in a climate that gets a decent amount of both sun and wind, but you're not sure either is enough to fully power your household, you could consider a hybrid system that uses both.

Geothermal Energy

As the components of its name indicate (geo = earth, thermal = heat), this form of power is dependent on heat obtained from the ground. This system works by pumping water through a pipe system that is buried underground. In the winter, the water in the pipes is heated by the warmth underground and then carried up to your heating pump, where that warmth is transferred to air that will be circulated throughout the house. Because it extracts heat from the ground, which is relatively warm, rather than from the winter air, which is very cold, this system is extremely efficient, with an input-output ratio of four to one. Unfortunately, you must have a professional install a geothermal heat pump, so the upfront costs are high. This type of energy system works especially well to heat and cool your home. It is less common than wind or solar energy, but as technology advances, it is becoming more common over time.

Hydroelectricity

This form of energy is only an option for those who have a source of running water, but if that's true for you, it's a great way to generate energy. These systems work by placing a mechanical component, usually a turbine, into the water and generating power from the turning of that component. Running water is much more constant than wind or sun, so the supply of energy from a micro-hydro system is more regular and therefore dependable than those systems. Hydroelectric systems need to be consciously designed and constructed so that they don't disrupt the natural environment of the water source, most likely requiring professional installation, and like wind turbines, they will require routine maintenance and upkeep.

Generator

Many people are familiar with generators, and some may even already own them as a backup power source for when there are disruptions to grid power. Generators can also be used as a backup for your renewable energy sources, and are especially recommended for any areas that are prone to tornadoes or other weather that might knock out power. Generators can run on gas, diesel, propane, or occasionally other fuels. When choosing a generator, it's important to get the right size if you want it to be able to power your whole household in the event of a failure from your primary power source.

Energy Storage

It might be strange to think of storing energy the way that you would food or water, but when you're generating your own power, any amount you can store in the present could end up helping you in the future. One method of storing power is using your local public power utility. Most utilities will store your power for you (and if you're sure you won't need the extra power, many will also buy it). This, of course, means that you would have to remain connected to the power grid. The other option is to store your excess energy in batteries. Batteries designed to store power like this come with a high upfront cost, but should last at least five years, and even longer if you care for them properly.

Waste Disposal

Setting up your own waste system might seem intimidating at first, but don't let that deter you. Dealing with waste is a problem many homesteaders have confronted and solved, and there are a wide variety of options available. Whether you're prioritizing ease and comfort or sustainability and simplicity, there is a waste disposal system that is right for you.

Regular Plumbing System to a Septic Tank or System

This method is both common and easy to get approved by your local authorities. Your inside system would consist of toilets and pipes, just like most regular plumbing systems. The waste from this system then drains into a tank or septic field on your property. Once in the tank, solid waste will move to the bottom to be broken down by bacteria, while the water will evaporate out through a specially designed pipe. Regular maintenance of a septic tank involves having it pumped out every few years. This will remove the solids at the bottom that haven't broken down (called sludge) as well as the grease and fat that has floated to the top (called scum). Pumping empties out the tank so it is ready to start the process over again. It is important to be aware of how the weather affects septic systems, especially in cold climates. If there's a risk of freezing (which can crack pipes and tanks), then it is best to have an underground system or one that is very well-insulated.

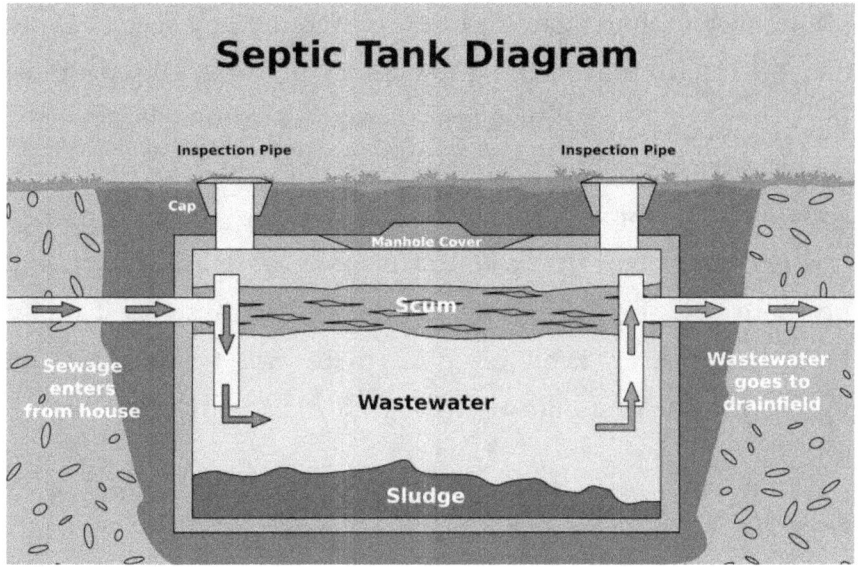

DIY Septic System

This is similar to the above, but you would do all the design and installation work yourself. As homeowners, you are already responsible for getting the correct permits and following all codes, so some people prefer to take the task on themselves rather than pay someone to do it. While this requires a lot of research and hard work, it also means you thoroughly understand your septic system and are fully equipped to take care of maintenance or any problems that arise.

Outhouse or Pit Latrine

On the other end of the spectrum is an outhouse, which is essentially a hole dug in the ground with a cover or seat over it and a small shelter around it. You're probably familiar with the idea of outhouses historically before indoor plumbing was developed, but many living on homesteads or off the grid still use them. Building a latrine does not require any special skills, but it is important to keep a few things in mind. Perhaps most important is the location, as you'll want to put it well above the waterline (especially if you have a well), a good distance away and downhill from any water sources, and above the flood level. It is also important that you only dispose of organic waste, and not pour any sort of chemicals into the outhouse, as it might react badly. Also, keep in mind there will be a release of methane gas (along with its strong odor) as your waste breaks down.

Honey Bucket

The honey bucket is essentially the small, portable version of an outhouse, consisting of a plastic bag inside a container, topped with a seat. This method is not highly recommended, because once the bag is full you have to dispose of the waste yourself, either by emptying the bags into a hole you dig or finding a local waste center that accepts human waste. However, it will work if you find yourself in need of a portable toilet alternative, and it can be handy to keep as a backup since it is small and easily stored.

Compost Toilet

This method of waste disposal will appeal to anyone who wants to do more with waste than simply get rid of it. Like an outhouse, the construction of a composting toilet is fairly simple. A container, usually a bucket, is placed under a specially-made toilet seat. After using the toilet, you add a layer of sawdust to soak up the moisture. There are also versions with electric or hand-turned cranks that speed up the composting process. Once the container is full, it can be added to your compost pile. This method not only keeps your compost pile full but also saves water because it does not involve flushing. Most compost toilets also have vent hoses that keep the toilet from smelling. Compost toilets are only legal in some states, however, so be sure to check your local laws.

Incinerator Toilet

An incinerator toilet is powered by either propane or electricity, and it does just what its name implies—it burns your waste. If you don't want to compost your waste or dig a hole, but you'd prefer not to have a full septic system, this can be a good option for you. While you will save water, you will be required to buy paper liners for the toilet bowl. A downside of these toilets is the amount of energy that they use. Whether your model runs on propane or electricity, a steady supply of power is needed. For this reason, some manufacturers of incinerator toilets recommend not using solar power or other methods of power that can be inconsistent or unpredictable with their products.

Lagoons

Septic systems work by filtering waste through the soil, but in some places, the soil will not work well for this purpose. In that case, you might need a lagoon instead. A lagoon works by collecting your wastewater in a (usually man-made) depression in the ground, where it can then be broken down by microbes. Even if your soil is fine, a lagoon can be a cheaper alternative to a septic system. However, not all areas allow for lagoons, and some people will be put off by having an uncovered pool of waste on their property.

Greywater Systems

Wastewater is actually broken into two different types, greywater and blackwater. Most of what we think of as waste is usually blackwater, meaning it is contaminated by human waste. However, a significant portion of wastewater is actually greywater, which is water that has gone through the system—think showers, sinks, washing machines, etc.—but is not contaminated by human waste. This water can be reused for certain tasks where the purity of the water doesn't need to be held to strict standards, such as watering your garden or flushing your toilet.

If you'd like to divert greywater so that it can be reused like this, there are several methods of doing so. Some people simply collect used water in buckets by hand to be reused, while others install plumbing that reroutes their greywater. Most frequently, this plumbing will take water draining from places like your washing machine and bathtub, and redirect it to one of two places: your garden irrigation system to water your plants, or your bathroom plumbing to be used to flush your toilets. In general, by using greywater instead of freshwater to flush toilets, you can reduce your water consumption by around a third. However, if you intend to use greywater in your garden, be sure that any cleaners you are using (like soap, shampoo, and laundry detergent) do *not* have any chemicals in them that will harm your plants.

Trash Removal

If you're located in an urban or suburban environment, you're likely going to continue to use the municipal trash services provided by the local government. However, if you're somewhere more rural, you might be in charge of your own garbage removal. This is another area where learning to reduce and reuse will come in handy. Everything that you can avoid (like plastic bags) or reuse (like containers) is something that you don't have to haul when you get rid of your trash. Next, be sure you're separating out things you can compost and recycle into separate bins from your regular trash. (I'd recommend a smaller compost bin so it can be easily transferred to your compost pile when full without too much trouble.) Recyclables can be taken to your nearest recycling center—check with your local authorities for locations and guidelines on what can be recycled. Everything that's left will need to be taken to the nearest landfill, with the exception of anything that contains hazardous materials—again, check local regulations.

Heating

Last in our exploration of utilities, but certainly not the least important (just ask anyone who's ever spent the winter in a drafty farmhouse!), heating is perhaps the utility most affected by lifestyle and planning. The amount of energy you exert heating your house is directly related to your house's space and insulation. A larger house means more space that must be heated, and a drafty house with lots of leaks where hot air can escape and cold air can sneak in will require more energy to heat. So if you're planning to build a house, be sure to include proper insulation and sealants, and consider eliminating square footage that you don't need. If you're working with a house that already exists, it could be worth the time and effort to check for drafts and seal and insulate where you can, as best you can. Gaps can be filled with sealants or foam; thin windows can be reinforced with plastic sheeting or covered with heavy shades; empty rooms can be closed off. You can also consider an exterior windbreak to block north winds. All of this can help make whichever of the below methods you choose that much more efficient.

Wood-Burning Stoves

Unlike fireplaces, which are open to the air through the chimney and therefore tend to be inefficient, wood stoves can work well to heat your house. Their upfront cost is reasonable, and depending on your land, you might have access to all the firewood you need. If not, firewood is relatively inexpensive and easy to get. If you'd like to make your wood stove work even better, you can keep a few weeks' worth of firewood inside, allowing it to warm up first and therefore take less energy to burn. You'll need a method, such as fans, to circulate the hot air they create. Most importantly, this method of heating needs constant supervision and maintenance.

Propane Heaters

These types of heaters can be portable or they can be built-in. Since they run on propane, they don't require electricity. However, you *will* need to be sure they are properly ventilated, including possibly opening a window at times to manage the fumes. Propane degrades very slowly, which means it can be stored for a long time—however because tanks are pressurized, they must be stored carefully and appropriately. A lot of propane heaters will have thermostats, which makes regulating the temperature (and therefore the amount of energy you use) very easy.

Solar Heat

There are two ways to heat your home with solar energy—passive and active. Passive solar heating means you take advantage of the heat naturally created by the sun's rays. This could be as simple as having wide, south-facing windows that let in a lot of warm sunshine, or it could be a more complex setup involving thermal walls or water tanks to amplify the sun's warmth. Even a wall made of dark material or painted in a dark color can help to absorb and then radiate heat from sunlight.

Active solar energy is more in line with general solar power systems. There are methods to collect warmth from the sun and then redistribute that through the rest of the house. A common way of doing this is to place water in conductive containers, such as glass or copper, and then pipe that water throughout the house once the water is hot. If you've ever lived in a house with radiant heat, the concept is very similar—but in this case, it is not a water heater but the sun that is heating the water.

Biomass System

When biological materials begin to decompose and break down, that process creates heat. Biomass systems work by placing metal pipes inside of piles of compost and other degradable waste. Those pipes absorb the heat from the pile and then redistribute it to other locations. This type of system probably cannot generate enough heat for your entire household, but it can be a handy and environmentally-friendly way to complement your main source of heat. You'll already be generating the waste—you might as well use it!

Next!

Hopefully, by this point, you've got a good grasp on what types of systems you might like to try in your own homestead. As I'll keep reminding you, there are *so* many options available, and so many ways to implement those options to fit a variety of lifestyles. Now that you know what choices are available for your utilities, the next chapter will walk you through another huge part of homesteading—growing food and raising livestock.

4

FARMING AND RAISING ANIMALS

A big part of creating a more independent life for yourself through homesteading is gaining the ability to feed yourself. Imagine never having to make another grocery run, and never wondering what, exactly, is in the food that you eat. With some hard work and careful planning, you can make that happen—and this chapter will help you!

For your first step down this sustainable food path, I recommend planting a garden. While some people eventually run large farms with acres of crops that they sell for a profit, for most homesteaders, planting fruits and vegetables is all about providing for their families. For this purpose, a large garden will be more than enough. The key is to choose what you plant carefully; always consider the nutritional value of your garden's yield. Everyone needs protein, carbohydrates, and fats in their diet, as well as vitamins and minerals. The goal is to grow enough variety that you can get a balanced diet from your harvest and to accomplish this, planning is essential.

Planning Your Garden

Careful planning before you begin your garden can save you a lot of time and money, not only by avoiding costly mistakes but also by helping you optimize your space so that your garden is as productive as possible. The first thing to consider is the location. You need somewhere with ample access to sunlight and water—beyond that, you can tailor your garden to fit your situation and needs. You can either start with a location and let its characteristics dictate how you'll set your garden up, or you can make a list of your ideal garden traits and look for a space that will fit them. If you don't have land or a yard, you can start with a container garden, such as a vertical garden, potted plants, or window boxes.

Once you know where your garden will be, you should collect some data. Measure the area and if possible draw a plan of where you'll be putting each of your plants. Also, track how much sunlight the area gets, and if you're planting in the ground, testing the soil is a good idea. Dig a small hole and examine the soil you take out of it; the best soil is porous and not too hard, able to be crumbled by hand into smaller bits that keep shape under pressure. Test the compactness of your soil by pushing a thin wire down into the earth—you should be able to get about a foot down before the wire bends. You can also look for what lives in the soil as an indication of its health—earthworms are a particularly good sign. Lastly, you can get a testing kit that will tell you the pH balance of your soil. Different plants will do best in different acidities, but generally, you want a pH between 6 and 7.5.

Once you've gathered all of this data, move on to supplies. I recommend making a list of all the supplies you'll need to get started—some possibilities are stakes, a watering system, a book for your notes, and containers if you're planting a container garden (though you can also make containers yourself). Especially when you start your garden, before things start to grow, you'll want to be *very* organized so that you know where you've planted each plant.

When it comes to planning the actual contents of your garden, a number of things will determine which plants are most appropriate. Consider your climate, as certain plants can only grow in certain climate zones. The amount of space you have is also an important factor, as some plants require a lot of space while others do not. Beans, for instance, usually need half a foot or less between plants; large gourds, on the other hand, can need as much as four feet.

Also be aware that certain plants grow well with each other, while others do not. For example, basil helps tomatoes and peppers grow by keeping away insects, while mint, chives, and garlic will keep slugs and aphids off your lettuce. On the other hand, keeping tomatoes and corn apart will stop them from sharing diseases common to both plants, and keeping lettuce away from parsley will stop the latter from crowding the former. These are just a few common examples, so be sure to research the specific interactions of the plants you're considering before purchasing seeds or seedlings. Another thing to be aware of is plant height. Taller plants should be placed farthest from the direction of the sun so that they don't block the light from other plants.

Timing is also an important factor of planning, especially if you intend to preserve what you harvest. Make sure you space out your harvest times. It's best to do all this planning in winter or early spring so that once the weather is warm enough, you can jump right into planting. While you're waiting for the weather, you can start seedlings inside.

Garden Irrigation Systems

If you're going to have a large garden, you might need to create a dedicated watering system. There are three basic types of irrigation systems that you can use to water your plants: soaking, dripping, and spraying.

Soaking uses a specially constructed hose that has many tiny holes in its surface that allow the water to seep out slowly. The hose lays beside your plants and the water seeping out saturates the ground around their roots.

Dripping follows a similar principle, but instead of tiny holes all over the surface of the hose, the holes are a little larger and set at intervals. You line up these holes to deliver water directly to specific plants.

Spraying utilizes sprinklers to deliver water over large areas.

Each system has its own benefits. Determine which to use based on the area that you need to cover, the amount of control you'd like over where the water goes, and the specific needs of your plants. For example, you might like a soaking irrigation system for plants that need frequent watering but whose leaves need to stay dry, versus a spraying system that will cover all parts of the plants with water. Or you might choose a dripping irrigation system if you only want to water certain plants at a time. You can also combine any of these systems with regular garden hoses, using the regular hose as a connector for spaces between plants where you don't want to waste water.

Raised Beds

A raised bed garden elevates the area where you're planting, whether through containers or built-up mounds of earth. This type of garden can have several benefits, including the ability to choose the type of soil you use, and bringing up the height of your plants so they are easier to tend and harvest. It can also keep the roots of your plants away from potential contaminants in the ground. The drawbacks of raised beds are their cost and the higher level of exposure, meaning your soil will be more susceptible to heat, cold, and dryness. You also need rows between the beds to walk through, so if you don't have a lot of space, this option might not be for you.

Designing a raised bed garden involves a lot of personal preference, but the average dimensions are 12 to 18 inches tall by 3 to 4 feet wide (length can be whatever works for you). If boxing in your beds, you can use wood (be sure any treatments on the wood are natural), concrete blocks, or metal. You may have seen old tires used as planters, but I don't recommend that; over time, the chemicals in the rubber can seep into your beds and harm your plants (and you, when you eat them).

Aquaponics

One of the more creative ways to grow food in a small space, aquaponics is the pairing of aquaculture and hydroponics. By combining these two things, you're creating an ecosystem where fish waste nourishes the plants and the plants' use of this waste keeps the water clean. Not only is this less work for you, but it also helps you avoid using commercial fertilizers that can have harmful chemicals in them. Because an aquaponic system raises fish *and* grows plants, it is a great way to provide a lot of food with a good variety of nutrients without needing much space. You can grow almost any type of plant with an aquaponics system, but some plants are particularly well-suited to it, and a few will give you a hard time. Tomatoes, peppers, lettuce, cabbage, cauliflower, ginger, basil, and strawberries will all thrive in this setting. You'll want to avoid blueberries (they tend to need a lower pH than other plants), chrysanthemums (these need a higher pH), and mint (which will grow well, but so quickly that it can choke out other plants).

You can make a simple aquaponics system by adding plants on floating mats to the top of a fish tank, or you can make a more elaborate system containing multiple containers and more hardy plant support. Aquaponics allows you to start small and build up your system as you become more experienced. Whatever the size of your setup, you'll need a pumping system that will help circulate the water between the fish tank and the plant beds. You'll also need a growing medium for your plants and the bacteria you'll add to help maintain balance. Aquaponics does not use soil, but rather materials such as sand, clay balls, and gravel.

An illustration of a basic aquaponics system

The plants and fish you choose for your system will depend on your location, the size of your tanks, and what type of temperature exposure you can expect. You want to place the system somewhere that gets a lot of sun, but you might also want a screen or other material to shade the fish tank from direct sunlight so it doesn't get too hot. Since you won't be able to move things once they're running without a lot of work, make sure you're completely confident of your location before you start to build. And because the system needs to be carefully calibrated and all the chemicals in the water balanced, you'll want to do more research and consult experts if you're planning to set the system up yourself.

Having a Root Cellar

Once you have all this delicious food grown and harvested, you're going to need somewhere to keep it. A root cellar is a great way to store food off the grid without the need for refrigeration. Root cellars work by taking advantage of the naturally cooler temperatures underground. The temperature of the soil at ten feet underground is generally around 50 to 60 degrees Fahrenheit, and the earth around and above your root cellar will cocoon it, acting as natural insulation to keep the temperature stable.

Your root cellar can be as small or large as you'd like. The smallest of root cellars are just containers that you bury. This method is simple and easy, and only takes the time you need to dig a hole. However, it is obviously more suited to storing foods that you want to keep for later, but won't necessarily need to access regularly. The container should be made of material that is waterproof and strong enough not to collapse or degrade over time, so metal and plastic are your best bet.

For larger root cellars, if you're lucky enough to have a dry, unheated basement with space to share, you can convert part of it into an indoor root cellar. Otherwise, you'll need to dig somewhere on your property. You first have to dig out space, then reinforce the floor and walls. A concrete floor is best if your budget can handle it; if not, you can use stones or wooden planks. It's important to have a floor to give yourself a level space that can support weight, insulate the temperature, and discourage burrowing animals from trying to get into your cellar. You can use stones, bricks, or masonry blocks for the walls, though the latter two are a bit more expensive. Finish off your cellar by capping it with a concrete slab if you can, and making sure your door and other access points seal well to keep out unwanted pests.

Once you have your root cellar constructed, you have to add a way for the air to circulate. Some of the foods you store will release ethylene gas as they ripen, and you need to have ways for that gas to escape and for fresh air to come in. At this point, you also need to decide what foods you'd like to store, as this will determine exactly how your root cellar functions. Foods with a long shelf life often do better in a dry environment, while fresh produce needs some moisture. If you plan to store both in your cellar, you should either create separate sections or seal your dry foods in airtight containers. You also should keep any items that might rust, such as cans, in sealed containers.

It's a good idea to outfit your root cellar with shelves, as that will allow you to keep your food off the floor and store your foods at different elevations to cater to their needs. For instance, you can keep foods most likely to produce ethylene gas on the top shelves, as the gas will rise and stay away from other foods. The space closer to the floor will be cooler, so you can keep any foods that need a little extra chill on your bottom shelves.

The last thing to remember about storing your foods is to know and track their expiration dates so that you can eat foods before they go bad. With a little bit of organization, you can have a system that keeps you consuming ripe food and replacing it with fresh food. You should also routinely look over the cellar to be sure all your seals are intact, you have no rodent problems, and the temperature and humidity are where you want them.

Foods to Stock Up On

The best foods to keep on hand, whether purchased or grown, are foods that have a lot of nutritional value and will last for a long time. While certainly not extensive, here is a list of foods with a long shelf life to get you started:

- brown rice
- oatmeal
- granola
- beans
- nuts (dried, unshelled, unsalted; in particular, pecans, pistachios, and sunflower seeds)
- peanut butter
- canned fish (in particular, mackerel, sardines, and tuna)
- beef jerky
- apple sauce
- dried fruit
- honey
- powdered milk
- oil (specifically coconut oil)
- spices

These will obviously be supplemented by what you grow yourself. In terms of meat, rabbits and poultry will not take too long to raise to adulthood (two to three years) and poultry will provide you with eggs in the meantime. For crops, winter squash and beans can be harvested the first year they're grown. Some other staples that will take longer but are worth space in your long-term plans include corn, potatoes, apples, and wheat if you intend to eventually be completely self-sufficient.

Livestock

Raising animals can be a huge part of your food self-sufficiency. Whether you dream of a farm full of different kinds of animals or just a few chickens in a tiny yard, adding livestock to your homestead can have many benefits. Your animals can be a source of fresh and organic meat, eggs, and milk. With milk, you can also make dairy products like butter, yogurt, cheese, and even ice cream. Grazing animals will also help keep your grass short, and animal waste can be used as a fertilizer for your garden. Caring for animals can be a great way to involve your whole family in the process of homesteading. And, if you intend to breed your animals, you can add those offspring to your own livestock, sell them for profit, or trade them to someone else within the homesteading community for other things you need.

This next section will take you through some of the most common animals you might want to keep, and the pros and cons of each. Something to keep in mind, no matter what type of animal you plan on raising, is that all livestock require regular health checkups. When you first get your animals, you should be sure they have had all required vaccinations and blood tests and have been dewormed. The specifics of these will vary by animal and location. For example, most mammals will need vaccinations for tetanus and rabies, and many states require birds to be tested for *Salmonella* and avian influenza. Your large animal veterinarian can let you know what type of regular care your animals need, and if you're new to the area and don't have a veterinarian yet, you can ask your local homesteading community for recommendations.

Chickens

Chickens obviously have the advantage of being quite small, and therefore easy to raise in a lot of different environments. If you plan to raise chickens in an urban environment, however, be sure to check the local laws. Many cities will allow you to have chickens but not roosters. Chickens can live free-range (allowed to roam as they please), in a coop with a yard, or in specially designed chicken cages. Even if you choose to cage your chickens, it is good to have enough room for them to move around a little. Each chicken should have a minimum of two square feet and enough height that they can easily walk around upright. About six hens will be enough to have eggs for the average family. You can also raise chickens for their meat. You can either keep a different set of chickens specifically for this purpose, or you can raise female chicks to lay eggs and male chicks for their meat.

For the healthiest chickens (and therefore the best amount of eggs), feed them both grain and protein-rich chicken feed. They will also eat your leftover food, grass, and bugs that they find while grazing. In fact, chickens are great at eliminating pests. They will peck away at your garden, so be sure to keep them separate from that—unless you are just starting your garden, in which case you can let them loose to tear up the ground, making your job of planting even easier. In the same way, you can let them into your compost pile, where they will tear up most organic material, helping it to decompose faster (as well as sometimes adding their own waste to the pile).

How often chickens lay eggs is dependent on a number of factors, including nutrition and weather, but the most prominent factor is the amount of daylight they receive. Usually, chickens will lay an egg daily during spring, summer, and fall, when there is a longer period from dawn until dusk. Once the hours of light per day dip below twelve, chickens are likely to stop laying. This is not true for all chickens but will be the case for the majority. If you really want to extend the laying season, it might be possible by introducing artificial daylight into your chicken's habitat, but be sure to talk to a professional about the potential impact that this may have on your chickens' health before deciding to do so.

Geese and Ducks

Both geese and duck eggs are larger than chicken eggs, though they also tend to stop laying once the weather gets colder and there is less daylight. Geese require a bit more space than chickens, given that they prefer to roam and graze for their food. You can also feed them grain at night. One warning about geese is that they're very popular with foxes, so they need to be kept inside a shelter at night. Rats, though perhaps not as big of a threat, can also be a problem, so be sure that your enclosure for your geese is rat-proof.

Ducks are susceptible to an even larger number of predators, including skunks, raccoons, coyotes, owls, hawks, snakes, and even cats and dogs. It is important that your duck shelter is off the ground just a little, not only so it is more secure from these many predators, but also so it stays dry. Also, ensure that their shelter contains enough soft materials (such as straw) for the ducks to make nests.

You might think ducks need water because we associate them with it so strongly, but ducks only need enough water to drink. However, having a source of water to swim in will make your ducks happier, so if you're able to provide that, you should! Ducks can be fed from a feeder, as long as you're sure to keep their food and water separate. They are also happy to eat weeds and bugs if you have the space to let them roam during the day. In fact, they will eat bugs off of your garden, and won't damage the plants as chickens would; just wait until after your plants are tall enough that the ducks won't accidentally step on them.

If you are planning to raise ducks and chickens together, you have to be careful with your male ducks (drakes) and female chickens. Drakes are not discriminant and can try to mate with chickens, but the two are not compatible, and your hens may be injured. However, if you have enough female ducks (usually three per every male), then the drake should be satisfied by mating with the other ducks and leave your chickens alone.

Rabbits

While they, unfortunately, don't lay eggs, rabbits have a lot of other advantages, particularly if you're new to livestock. They are fairly docile and don't make a lot of noise, and they are easy to breed yourself if that's an interest. Rabbits are happy to consume garden leftovers and any other greens that you don't want, and raising rabbits can also help your garden grow, as their waste has good fertilizing properties.

Rabbit meat is also very healthy for you, as it is quite lean and nutritious. If you're planning to butcher your animals yourself, rabbits have a much nicer learning curve than some other animals, especially those with feathers. You can also use the rabbit hides to make things, and certain breeds (like the Angora) have fur that you can shave and make into yarn. You can use the things you make from pelts and yarn around your own house, or you can sell them for additional income.

Pigs

If you're planning to be food independent while also having bacon, you'll need to raise pigs. Some people prefer to breed their own pigs, thereby supplying their own stock, while others would rather buy piglets each year and then raise them to adulthood. The choice is up to you. When buying pigs, be sure the farm or breeder has a good reputation for healthy animals. While it may be tempting to go for chubby piglets, that can actually be an indication that the adult pig will carry a lot of lard. It's also good to know that the longer a pig has been nursed, the healthier it will be.

Pigs will eat leftovers from both your table and your garden. This helps keep the cost of feeding them low, although it is good to be sure they get some grain. If you have the room, you can supply this yourself by raising a small amount of corn. You can also give your pigs mineral and protein supplements to keep them healthy. Pigs will not overeat and can be fed from a feeder, meaning a large amount of their care can be automated. When saving scraps for your pigs, be sure to separate out anything that could harm them, such as paper, plastic, or anything with chemicals.

For living space, you can get your pigs a hog house or build them a shed. Make sure that the floor of this space is high enough off the ground that it will stay dry, and that sunlight can get in to help keep it fresh. It is also good to keep in mind that when piglets are very small they can slip through many spaces, so plan your pen accordingly. The smaller the space, the less exercise the pigs will get—and therefore the quicker they will grow. But you will want to find the right balance because cramped spaces can become unsanitary and allow parasites to grow on the pigs.

When harvesting pigs for meat, you can do it all yourself, have someone else do the entire process (from slaughter through curing and smoking), or you can have someone slaughter and dress them before returning the meat to you to do your own curing and smoking. You can check with neighbors, breeders, or feed suppliers for recommendations about where to get your pigs slaughtered.

Sheep

As their diet consists mainly of grass, sheep are fairly inexpensive to raise after the initial cost of purchase. Depending on the quality of your land, you will need about an acre of land for every two to three ewes (plus lambs). If you have the room on your property, you can start them at one spot and move them once they've eaten most of the grass. (Always graze sheep after cows, never cows after sheep, as the sheep will have eaten too much for the cows to find food.)

It is best to look for a breed that is native to your area. Starting with a small number of ewes is recommended. If you want to breed your ewes, you can add a ram, or you could look into borrowing a ram for breeding. You can expect each ewe to have one or two lambs on average.

Goats

Goats are great for those who are just starting out with livestock. They take up less land and need less food than cows, but they still can provide both milk and meat. Goats will eat all kinds of plants, including herbs, trees, and bushes, which means you can let them graze for a large portion of their diet and they can be helpful in clearing land, eating the brush that you would otherwise have to pull up. They don't require a lot of supervision, though they are known to be smart enough to escape their pens if given the chance. Therefore, it's important that you be sure their enclosure is escape-proof.

For many people, goat milk is easier on the stomach than cow milk. It is a preferred alternative among those with digestion problems, including young children, older people, and people with ulcers. The American Dairy Goat Association has guides on eight different breeds of dairy goats; among those, Nubian goats are often considered the best. They are great milk producers, and their milk has a high amount of both butterfat and protein.

Goats are a good source of lean red meat. Traditionally, cows have been a bigger source of red meat in the United States, but goat meat production has been growing since the early 1990s. Some of the main meat goat breeds are Spanish (wild) goats, Boer goats, Kiko goats, Myotonic (fainting) goats, Savanna goats, and Texmaster goats. Generally, the Boer breed, brought to the U.S. from South Africa, is considered to have the best meat.

Cows

The largest animal on this list, cows are also the largest commitment in terms of both money and time. Dairy cows must be milked multiple times every day, and both dairy cows and beef cattle require a significant amount of food and land. However, cows also offer a big return on investment, from meat to dairy products to offspring that you can sell or trade. There are hundreds of different breeds of cattle, but these are some of the most popular that you might consider for your farm:

Angus — This breed and the beef it produces are very well known in the United States. If you're planning to raise beef cattle, this is definitely one to consider—especially if you plan to sell what you raise, given the breed's well-known brand.

Hereford — These cows are mainly raised for beef, though they are sometimes also used as dairy cows. They are found all over the world because of their ability to adapt to a variety of climates, so if you're hoping to raise cattle but are worried about their suitability to your local weather, you may want to look into these.

Charolais — As you might guess from the name, this type of cow was first found in France. These cows are well-known for the quality of their meat as well as their hides and are often cross-bred with other breeds.

Galloway — A breed from Scotland, these cows have distinct fluffy coats. They are one of the oldest breeds of cattle and another strong beef breed.

Limousin — Another French breed that is now raised in many places, the Limousin is generally bred for its meat. However, they also make great work cows.

Simmental — This Swiss breed has been in the United States for over 200 years, and while it is often used for dairy production in other places, here it is more commonly known as a beef breed. The Simmental is a fast grower and is one of the larger breeds of cows.

Shorthorn — The Shorthorn breed was actually created to fulfill both dairy and beef needs, so they are a great choice if you want flexibility, or if you're yet not sure whether beef or dairy is right for you. Some breeders have cultivated a specialty for either dairy or beef Shorthorns, though, so be sure to discuss this with your breeder before you buy from them.

Holstein Friesian — These cows produce the highest volume of milk among all the breeds. If your household consumes a lot of dairy products or if you're hoping to make and sell them, this breed is a great choice.

Brahman — This breed is ancient and is well-known for its hardiness. It is very resistant to diseases, parasites, and overheating, is able to withstand extreme climates, and can also survive long periods with little food.

Scottish Highland — Hailing from Scotland, obviously, you might have seen this long-haired cow on postcards featuring the Scottish Highlands that gave it its name. Like the Brahman, the Scottish Highland cow is also very robust. It is less susceptible to disease and is able to thrive in both hot and cold climate zones.

Bees

Bees may not be the first thing you think of when you think of livestock, but beekeeping is a growing interest among homesteaders and the general population. The obvious benefit is the ability to make your own honey, but you can also get the satisfaction of knowing that your bees are contributing to the local ecosystem through their hard work pollinating local flowers. Consuming local honey is also thought to help with allergies.

As with all these animals, it is important to do your research, especially since improper handling of bees can lead to the very unpleasant experience of being stung. There are a lot of resources for learning about bees, including many local beekeeping organizations. Local organizations are particularly good for beekeeping because the specifics of your location will influence how to best care for your bees.

Spring is the best time to start a hive, so plan ahead and do your research the year before. You'll then be in good shape to place your bee order in the winter. You can start with a standard package of bees, or you can start with what's called a nucleus colony. A nucleus colony comes with a queen who will lay eggs, worker bees, other bees of various ages, and food. This setup means you are not starting the hives from scratch, so your bees have a better chance to thrive. It also allows your hive to grow quickly. You'll want to feed your bees with sugar water in the beginning, to help them get acclimated; you won't need to do this forever, though, because as they become familiar with the area and more flowers begin to bloom for them to feed on, they will stop drinking the sugar water.

It is difficult to move a hive once it's established, so be sure to place it somewhere safe and easy to access. You'll want to periodically check in with your hive by opening it, and for this, you'll need a specific set of tools: protective clothing (leather gloves, a veil, and/or a jacket), a smoker, a bee brush, a frame lifter, a multipurpose hive tool, and a set of regular pliers. You can purchase these new or used, but if you get used equipment, make sure you clean it thoroughly before introducing your own bees. Always start your smoker *before* opening your hive, as this helps to discourage the bees from stinging. The best way to tell if your hive is healthy is to look at the queen; if she's laying eggs, then she is happy, and a happy and healthy queen means the rest of the hive will be happy and healthy, as well.

Next!

After establishing your location, utilities, crops, and livestock, your homestead will be shaping up pretty well! You're informed about how to raise livestock and grow food, how to heat and power your home, how to dispose of waste, and how to provide yourself with water. If you can get all those things lined up, you'll have a good solid foundation on which to build your homesteading dreams. In the next few chapters, we'll look at how to prepare yourself for both exceptional situations, like medical emergencies, and day-to-day life, like standard homestead maintenance.

5

MEDICAL CARE

While it is important for everyone everywhere to be prepared for medical emergencies, it is particularly crucial if you're planning to live in a more remote area. It's a good idea to know how to treat minor injuries yourself and to know what to do in the case of major injuries or other life-threatening situations. Due to travel time, terrain, and other impediments, access to emergency services and medical care can be limited in rural areas. This chapter will describe both preventative measures that you can take to avert disasters before they happen and methods to deal with those that are unpreventable.

There are two categories that most emergencies fall into: those that jeopardize your home, such as a tornado or flood, and those that don't, such as medical problems. The best mindset is to do everything you can to prepare for both kinds on your own, but have systems in place to utilize whatever outside resources might be available. Below are some tips to be as ready as possible for whatever life throws at you.

Preparing for a Crisis

Be aware of your surroundings.

This is not just a call for in-the-moment awareness, though you should have that, too. You need to be aware of what might happen—weather, accidents, natural hazards, etc.—as well as what services are available to help you. There will be dangers that are particular to your area. Look into the history of your location and talk to long-time residents. Find out what threats might exist from local wildlife, and learn about those animals' behavioral cues and what to do if you encounter them. Know how far away traditional means of medical help such as hospitals are, and how long it will take an ambulance or other transportation to get you there. Essentially, map out what potential perils exist, and then make plans for what you will do if they happen.

Remote doesn't mean disconnected.

Communication is vital for homesteaders for a variety of reasons. Weather alerts via a weather radio or from a website that tracks your local weather can help you know trouble is coming in time to brace for it. A standard radio tuned to a local news station can also warn you of potential disasters. Even if you plan to take care of minor injuries yourself, be sure to keep contact information for emergency services. Have several backup means of communication, such as a landline as well as a cell phone. If you live somewhere with bad cell reception, consider investing in a satellite phone. And very importantly, test all your communication methods routinely. You want to fix any problems that arise *before* an emergency, not during one.

Remember your community.

Even if you're living quite far from city limits, most people will still have some type of neighbors—if the lot you're looking to homestead on is so remote that you have nobody to turn to when calamity strikes, you might want to keep looking. Just as you should have contact information for emergency services, you should also gather contact information for any neighbors close enough to help you in an emergency. Don't only call them if something goes wrong, though—foster a relationship with them. Be willing to help them when they need it, which will build camaraderie and trust. Being there for one another is not just a sentimental idea for homesteaders—it's a way to guarantee everyone's survival.

Put together emergency kits.

Whether dealing with a small misfortune or a huge catastrophe, the last thing you want to be doing is running around collecting items that you need. The idea of a first aid kit is familiar to most people, but you should also pack bags that you can grab at a moment's notice. Fill these with essentials such as emergency cash, copies of important documents, a battery-powered or hand-crank radio, flashlights, extra batteries, a power bank for charging cell phones and other small electronics, matches (in a waterproof container), a whistle, wipes, soap, basic tools, small amounts of non-perishable food, bottled water, a can opener, eating utensils, maps, hand sanitizer, and medical supplies, including any medications you need. You might include additional things specific to your household, such as pet supplies, baby supplies, etc. Having these bags on hand will give you peace of mind, and make you less likely to panic when you need a clear state of mind.

Discuss your plans.

It is important to plan for any and all scenarios, but it is equally important that everyone in your household knows what those plans are. Routine discussion will help keep things fresh in everyone's minds and make sure that you are adapting to any changed circumstances. Be sure that you have a designated place where everyone will meet in the event that you must leave your home. You might even have several different points, in the case of different situations. You can also keep other emergency supplies at your meeting point, such as more food, blankets, extra clothes, larger containers of water, a fire extinguisher, a temporary shelter such as a tent, etc.

Level up your skills.

There are a number of courses you can take that will help you be more prepared and useful in an emergency, including first aid, CPR, wilderness survival, and wilderness first responder classes. If you're unable to find a course near you, try checking online. These will teach you to assess a situation and know how to proceed. You'll learn to treat small injuries and what to do to mitigate the damage from larger ones. Not only will these courses increase your expertise, but they will help you keep a level head in a time of crisis. It might be cliché to say that knowledge is power, but it's cliché because it's true. When you know what is happening and what can fix it, you will be empowered to act swiftly and decisively.

Common Injuries

The best way to minimize injuries is to anticipate them. If you assume that you will not get hurt, injuries will take you by surprise, and you'll be left with no clear way to handle them. However, if you assume that it's likely you will get hurt, not only will you be ready when it happens, but you'll also be training yourself to stay cognizant of potential harm. Here are some frequent injuries to be aware of and prepared for:

Blisters — Blisters are one of the most common skin irritations. Friction, exposure to extreme heat or cold, and allergic reactions to plants or insect bites all can cause blisters. Common sense can save you from many of these. Wear proper clothing that fits well and is appropriate to the weather, make sure your skin is covered when handling unfamiliar plants or other substances, and use bug repellent.

Cuts — Even in every day, non-homesteading life, we risk being cut all the time. Knives, garden tools, power tools, and even scissors can cut you if you're not careful. On a homestead, there are additional risks, such as using an axe or putting up barbed wire. Reduce your risk by wearing protective gear and always using the right tool for the task.

Muscle Injuries — Life on a homestead can involve significantly more physical activity than most other lifestyles. You'll find yourself lugging a lot of heavy things! This can be particularly stressful on your neck and back. Practice correct lifting and carrying posture, and never lift or carry something heavier than you can manage.

Heat Exhaustion — With increased physical exertion can come overheating, and you need to be especially mindful of this if you live in a warm environment. While it may seem more efficient to keep going, in reality, pushing past your limits can make you faint or dizzy. Take regular breaks and drink a lot of water whenever working in heat.

Accidents with Machinery — Chances are that if you live and work on a farm, at some point you'll need to use machinery such as a tractor or chainsaw. Due to the nature of these machines, injuries resulting from their misuse are often serious or even fatal. While there is no way to be completely accident-proof, many of these injuries happen because machinery is operated carelessly or in a way that it was not made to operate. Always be sure you follow all safety protocols and wear the proper safety gear, and keep all of your machinery well-maintained in good, working order.

Hunting Injuries — Obviously, hunting weapons can cause very serious or even fatal injuries. If you are going to be using a gun or other hunting weapon, the first step to safety is making sure that you are properly trained and certified. Always be sure to keep guns locked away from children and other untrained individuals. When you are hunting, wear the correct high visibility clothing, and be aware of your surroundings at all times to prevent accidental shootings.

First Aid Kit Checklist

You can obviously buy a first aid kit fairly easily, but you might want to make your own so that you can be sure it meets your specific needs. Try to get in the habit of replacing anything that you use right away, and go through your kit regularly (at least every six months) to replace anything you've missed or any of the medicines that have expired. Here's a basic list of what to keep in your first aid kit:

- Bandages, including one or two boxes of band-aids in assorted sizes, absorbent wound dressings, cloth tape, rolled gauze, a roller bandage, triangular bandages, and sterile gauze pads in assorted sizes
- An instant cold pack
- Antiseptic wipes
- Antibiotic and hydrocortisone ointments
- Aspirin
- A non-mercury, non-glass oral thermometer
- Tweezers
- Non-latex gloves
- A CPR mask
- An emergency (Mylar) blanket

Some other things that might be good to keep in or near your first aid kit are prescription medications or medications that you regularly take and important medical information such as the doctor's phone number, family contact numbers, health insurance information, and family member blood types.

Next!

While the information in this chapter may seem daunting, remember that part of being ready for emergency situations is considering and preparing for all possibilities. Not only will it keep you safe, but it will also help ease your mind. Knowing that you have the skills and resources to take care of yourself and your family in a crisis might even give you the confidence you never had when you were relying on others to do that for you.

6

HOMESTEAD MAINTENANCE

With homesteading, freedom and responsibility go hand in hand. You're in charge of things! Which means… you're in charge of things. And that includes all of the equipment that you'll use to make your homestead work. The last few chapters gave you an idea of the various systems, machines, gadgets, and tools that come together to keep a homestead running. This chapter is about maintaining all those things. As we briefly touched on when discussing accidents and medical emergencies, it is very important to make sure that everything is in good shape and working properly. Not only does that keep things safe, but it also improves productivity and helps you avoid unnecessary delays.

Organization

The first step to maintenance is to keep everything organized and to put things away when you're done with them. Tools that get left in random places will go missing or get damaged much more easily. My tip for keeping things neat and tidy is to group tools by type and have a designated container or area for each group. It doesn't matter how you categorize them—it could be by function (cutting, fastening, etc.), by application (plumbing, electrical, etc.), by size, or even by color if you want! All that matters is that you'll be able to remember where they go and why.

I also suggest keeping a list or chart with all of your more complex tools or gadgets that need regular maintenance, such as oil changes or filter changes. Record what you should check, how often, when you last serviced it, and the specifics of any part (such as batteries) that will need regular replacement. Keep this list somewhere easy to access near your tools so that you'll remember to update it as necessary.

Cleaning and Storage

This is common sense, but worth stressing anyway: keeping things clean is essential to keeping them in good working order. This is especially true for seasonal equipment that will be stored for large portions of the year. When you use something for the last time before putting it away for the season, make sure you *thoroughly* clean it. It should also be totally dry before being stored, as moisture can make things rust, rot, or grow mildew. Things that are used frequently should be cleaned and put away in the same spot after every use. Making sure equipment is stored somewhere away from bugs, dirt, and dust is important, especially if that equipment is used to handle food in any way.

Tool Maintenance

There are four big things that taking care of your tools prevents: dullness, rust, wear and tear, and stiffness. The first two are fairly straightforward. Sharp and rust-free tools are easier to use and work better, which also means they are safer. Some tools can be sharpened at home and others are best left to professionals—you should be able to get that done at a hardware store. Rust is easy to remove yourself with a little elbow grease and some steel wool, sandpaper, a wire brush, or a rust remover.

Wear and tear are often less about proactive maintenance and more about proper use, but you can help keep wooden tools in good shape by applying wood butter or oil to prevent drying, cracks, and splinters. As for stiffness, grease and oil are your friends. Grease in the gears of machines keeps them running well, and oil can be applied to any stuck joints, hinges, or other moving parts. Remember to be aware of how you use the tools and make absolutely sure that any grease applied to tools that might be used for or around food is non-toxic.

Vehicle and Machine Maintenance

As I suggested above, keep detailed records of your maintenance schedule for any complex machinery, including vehicles. Regularly inspect tires, hoses, and wiring, and change out batteries and fluids. Look for leaks, loose wires, and damage. Every time you finish using a vehicle, make sure the headlights and other parts of the vehicle that use battery power are turned off. This small habit can save you the immense frustration of an unexpected dead battery. Also, be mindful of how weather can affect vehicles. Heat and cold can alter the air pressure in your tires and gas can freeze if a vehicle is not frequently used in wintertime, so empty the gas tanks of any equipment you don't intend to use throughout the winter.

Maintenance Checklist

You can break your maintenance list into two categories: things to check often, and things that you should check a few times throughout the year.

Things to Check Often:

- Anything broken or worn out. This may seem obvious, but it is important to repair small problems as they happen to avoid the chance of them snowballing and becoming much bigger problems. Some common issues you'll want to fix right away are:
- Broken or blown light bulbs
- Water leaks
- Rusted metal
- Broken windows or other glass
- Chipped or peeling paint
- Cracked putty
- Traps or other forms of pest control. As with broken things, make sure to solve this problem right away, before it grows and you run the risk of infestation.
- Anywhere that water might leak. In addition to fixing

leaks quickly, it is important to do regular checks to look for leakage. Catching a leak early can be the difference between a minor inconvenience and a huge catastrophe.
- Your main electrical panel. There are a few things you should do here:
- Look for signs of moisture, such as watermarks or rust.
- Check all your breakers, turning them on and off to test that they're working.
- Tighten any loose fuses.
- Check for heat or the scent of burning. (If either of these is present, call your electrician ASAP.)
- Your emergency cash fund. This isn't necessarily a thing to do, but it is something to maintain. You should always have a bit of money stashed away for unexpected expenses, and if you spend any of that money, replace it as soon as you can.

Things to Check a Few Times a Year:

- Doors and windows. The biggest thing to look for here is drafts. Ensure that when they're closed, everything is sealed tight. You should also examine door and window frames for warping or instability, which could be a sign of bigger problems.
- Walls. It's likely you will notice most damage to your walls when it occurs, but it can never hurt to take a few hours once or twice a year and do a thorough check. If

you find cracks or holes, fill them and make note of where they are so you can make sure they don't increase. You should also look for water stains.
- Ceilings. Similar to walls, but you also need to check for sagging or changes in shape.
- Paint. Both interior and exterior paints provide a barrier between raw materials and everyday wear or, in the case of exteriors, weather. Repaint any blistered, cracked, or peeling areas.
- Pipes and fixtures. Make sure none of your taps are leaking. If you find a leak, try replacing the washer—that usually solves the problem. Check all seals on things like tile and grout, especially around areas like the toilet and the bathtub. Thoroughly investigate anywhere the seal might be broken as water leaks in these spaces can lead to mold, which can get out of hand if left unchecked.
- Roof(s). Look for damage, including loose shingles, warping, bubbling, or other indications of structural problems. Keep your roof clear of branches and other things that might have fallen onto it.
- Chimney(s). For wood-burning chimneys, call a chimney sweep annually to clean them out and routinely check for damage in the bricks or mortar. If you have a gas appliance in your chimney, have a qualified technician inspect it yearly.
- Steps, stairs, railings, etc. Check the security of *all* of your steps, both indoor and outdoor. Discovering a

wobbly railing or unsteady step through a routine check is vastly preferable to discovering it when it breaks or comes loose under your foot.
- Gutters and drains. Keep an eye on these all year round, but be sure to check and clean them of leaves and other obstructions before you expect periods of heavy rain. Drains can become loose over time and may need to be reset, so check this as well.
- Exterior wooden surfaces. Test the seal on your outdoor wood by dripping water on it. The water should bead up. If it soaks into the wood, then the seal needs to be redone. Be sure to sand the wood down before resealing it. Also, inspect any wood for signs of insect infestation or decay.
- Garages, barns, workshops, etc. Check anything from the above list that also applies to these structures—walls, ceilings, wood, paint, etc.
- Paved areas and/or driveways. Look for worn parts or cracks. Repair any uneven areas, especially if they are routinely driven on or if their slope might direct rainwater toward your house or other structures.

Next!

As we've made our way through the book, I've frequently reminded you that planning is essential to homesteading because careful planning can save you time, money, and headaches. Proper maintenance is an extension of that and is essential to making your homestead run smoothly. It will help you avoid costly repairs and delay-causing breakdowns while also just making your life easier. There is another essential organizational tool for your homesteading journey that we have yet to discuss, and that's your budget. The next chapter will take a deep dive into what a budget does, how to make one, and how to tighten it when you don't have a lot of funds.

7

HOMESTEAD BUDGETING

Why Is Budgeting Important?

You're probably familiar with budgeting, no matter your background. A budget is a good tool for anyone running a household of any kind. However, homesteading has a particular set of pros and cons when it comes to budgeting; there are many expenses that you likely aren't used to budgeting for, especially if you are working on a large scale with many additional vehicles or machines to maintain. Not only are these expenses unfamiliar to those used to urban or suburban living, but that equipment can be vital to the function and well-being of your household. Because of this, you need to be able to take care of these unexpected expenses in a timely manner. This means that you must have a cushion in your budget, so that one or two negative events don't mean severe setbacks or the collapse of your entire enterprise.

Another thing to consider about homesteading life is that it goes in seasons. There will be particular costs associated with each season (planting costs in spring, harvesting costs in fall, etc.), so your budget needs to be more than a predetermined monthly calculation of expenses and income. Similarly, your income likely won't be a set amount that you receive at a set interval. You might have a large amount of money come in at once, but you need to make that money last for a long time. A budget will help you know when you can spend money and when you need to save that money for later on in the year.

Right now, you might be working for an employer who is responsible for taking taxes out of your paycheck and paying them to the government. Much of your income in homesteading will not be for an employer—in fact, you might be employing other people. This means that you also need to adjust your budget to account for the taxes that you'll have to pay. And if you are transitioning from being employed to working for yourself, you have to account for the cost of that transition. You might be hoping to eventually make your homestead the main source of your income, but it will take time for you to turn a profit. This can be years, so it is very, very important to go into that endeavor with a solid budget, where you know what money you'll have coming in, how you'll be spending it, and generally how you'll survive until you are able to sustain yourself.

On the positive side, you'll find you have a lot more control over all areas of your life. This gives you the flexibility to choose less expensive options rather than being locked into whatever utility or service has the monopoly over your area. It also means you might be doing or making a lot of things yourself, eliminating the need to pay for them. All of this impacts your bottom line, and it's important to not just cross your fingers and hope everything adds up. Like the rest of this life, making your budget is about being willing to take on greater responsibility in exchange for greater control. The rest of this chapter is here to help you do just that.

How to Create a Budget

Step 1: Gather information.

This is the first step, but it is also one you'll need to be doing constantly, possibly forever. The more experience you gain, the more you'll know about what things will cost, how much money you can make from certain activities, and many other variables that will affect your budget. Smart money managers constantly take in new information and adjust their budgets accordingly. Every year, every season, even sometimes every day, you'll be updating your knowledge base. You can use that knowledge to make your household more financially efficient. This may sound like a lot of work, but once you've become practiced at it, it will become almost automatic.

But in the beginning, before you've got a lot of experience, gathering information will involve less personal observation and more research. You'll be doing most of this research already as you look into all the options that have been presented to you in this book so far. When you are deciding where to live, collect data about how much it costs to live there.

When you're choosing your utilities, make note of the prices of each system. When you're planning what plants and livestock you'll have, estimate how much each will cost you. As you anticipate what you'll need for maintenance and emergencies, record what the upfront expenses of your supplies will be, and how much repairs and replacements will be. Taking this small extra step for all of your planning will leave you with a great foundation of information with which to build your budget.

Step 2: List your expenses.

Start with the basics: regularly occurring expenses like food, clothes, phone service, and medicines. Next, account for monthly, quarterly, yearly, etc. bills; this includes any debt you're paying off, such as car loans, student loans, and credit cards. You probably have a pretty good idea of all of these already. From here, start adding in all the other information you've gathered, like upfront expenses, including equipment and livestock; seasonal expenses, such as new plants or animal feed; housing and living expenses, including your estimated mortgage or lease payments; maintenance and medical expenses; the upfront cost of your utilities, as well as any ongoing expenses for them; taxes you'll be responsible for; and anything else that you came across in your research.

This list is not exhaustive but is meant to help jumpstart things for you. Everyone's list of expenses will be slightly different because different households have different needs. The idea is to write down absolutely everything you can think of so that you are never surprised by an expense. Don't forget to include a cushion for unexpected costs—you might not know exactly how much this should be, but use your best estimating skills. You can fine-tune this number later. While you're making this list, also write down when you will need to pay each expense. It can be helpful to make a rough list with all your information to ensure you don't forget anything, and then redo the list more neatly, organizing things by date and category.

Step 3: Determine your monthly expenses.

It's not very useful to have one big list of all your expenses, especially when some are monthly, some are yearly, some are one-time, and some might not have a set date. So, the next step is to determine how much money you'll need each month. You want to spread out big expenses over the year—basically, you'll plan to save a certain amount each month, so that when those expenses need to be paid, you've accumulated enough money to pay them. To get the amount you should save each month, divide each yearly expense by 12. If an expense is set at a different interval, determine the yearly cost and divide by 12 (i.e., if you pay a bill four times a year, multiply by four and then divide by 12). If you're not sure exactly when you'll need to pay an expense, use your best judgment.

Note that you might have big expenses in the first year or so that you will not have time to save for if you start saving after you start your homestead. You can plan to pay for these with savings you already have, or you can start incorporating those expenses into your monthly budget now with the intention of building up your savings before you make the transition. (You could also take out a loan for those expenses, but be very careful with this option, given that you may not have the ability to start paying it back for a few years.)

Step 4: List your income streams.

Just as you need to know how much you will spend and save each month, you need to know how much income you can count on. If you have a regular income that you know will continue, start with that. Then, add in any income from one-time sources, irregular sources, or any savings that you're comfortable including. You can include projected income—meaning income that you *think* you will make, but don't know for sure—but be careful not to rely too much on this. Homesteading can be very unpredictable, especially in the first few years, and you don't want to be left in the red because you anticipated getting paid for something and it fell through. As with your expenses, write down when you expect to get each stream of income.

Step 5: Put it all together.

There are a few different ways that you can do this. Some people prefer to calculate their total yearly expected income, and then divide that by 12 to get an estimated monthly income. After the division, you have a simple subtraction problem, money in minus money out. Some people do a similar method with quarters (three months at a time). The guiding factor here is what makes sense to *you*. However you divide things, the goal is to be able to predict the money you'll be spending for a certain period, and distribute the money you'll be making so that it covers that spending. For example, let's say that you know you'll make $6,000 in June, and your budget says you have $3,000 per month in expenses. This means that the money you make in June will cover your expenses for July and August, but you will need to make more income for your September expenses.

If creating your own system seems intimidating, don't stress; you can always gather all your expenses and income, and then input them into a premade budget. Apps like Excel often come preloaded with budget templates. You can also find templates online for both Excel and other programs like Google Sheets. If you're really feeling out of your depth, you could also consult with a professional accountant. However you do it, the important thing is that you create a budget that works and you stick to it.

Homesteading on a Limited Budget

The last part of this chapter will focus on tips for getting started when you don't have much money to work with. Yes, there are some parts of homesteading that can be very expensive. But even if you don't have the funds for big upfront expenses or a lot of land, you can still do this. In fact, switching to a homesteading mindset can actually help you save money in a lot of ways! Making food and clothing for yourself, being focused on reducing and reusing, and just generally being self-sufficient can cut down on your expenses and help you get more in control of your finances. Read on for advice on how to get started.

Tip 1: Downsize your life.

Many people's mindset, especially in the United States, is that more is always better. We like to do more and own more and be "more," but constantly accumulating things has a point of diminishing returns after which more stuff just means more to take care of, more to pay for, and for whoever's doing the housekeeping, more to clean. Take a hard look at what you're spending money on and ask yourself if those things *truly* make you happy. If that sounds daunting, try thinking of it in terms of the direction you want your life to go. If you focus on your goals and the purpose you want your life to have, then you have an inherent guideline for what's important in your life. Examine how you're spending your money, and ask yourself, "Is this taking me in the direction I want to go?"

I'm not advocating for you to eliminate anything you truly love, but so often we're spending money on things that we're only guessing will make us happy. Stop guessing, and start intentionally choosing your priorities based on where you want to go next. A good way to practice intentional purchasing is to ask yourself if something is a need or a want, and then only buy it if you need it. Classifying purchases into needs and wants will help you streamline your expenses, and this kind of focused buying will have the added benefit of helping you clarify and solidify your goals and plans.

Tip 2: Be patient.

The best way to change your life is to take small steps consistently. You've decided you want to change your life—great!

That doesn't mean you should overhaul everything at once. If you're at the beginning of this journey, you'll likely have a lot of things you want to change. You can use those things as a guide, but remember the SMART goals we discussed in chapter one, especially the "A" for attainable. Take your big, long-term vision and break it down into very small parts that you can achieve a bit at a time. Make a list of all the things that you want to achieve, and then separate your list into things you can do now and things that you can't do yet. For the things that you can do now, make a plan on how you'll slowly integrate them into your life. For the things that you can't do yet, pick one or two that are top priorities for you and make another list of what needs to happen for those things to be possible. You can also separate that list into "now" and "not yet." Keep working like this until you have a clear plan of small steps that you can take now which can build into long-term change.

Tip 3: Start with what you have.

This covers a lot of the same ground as the last tip, but from a slightly different perspective. It's great for when you're feeling frustrated and far away from your goals, and it involves starting up your imagination and getting innovative. Take stock of your life and consider what parts of it can be adapted right now. Can you grow plants on your balcony? Can you start learning to sew? Do you have room for one or two small animals in your backyard? A few other ways to ask the same question: What skills can you acquire now that will serve you well in the future? If you're hoping to move eventually, what things can you build now that you can transplant to a new home when you're ready? This also dovetails into the first tip: while you're simplifying your life, look for things that you can save and repurpose. Containers, clothing, even old furniture all have possibilities of being reused as a part of your new lifestyle.

Expand this thinking beyond just your home. I've mentioned community gardens—find out if there's one near you. If you know someone with plenty of land, ask if you could use a small portion of it for a garden. Inquire at your local home and garden stores or nurseries about what happens to their leftover seeds at the end of planting season. Often, they will let you take them because they would otherwise be thrown away. If you learn to grow your own plants from parent plants, you can ask friends and family for clippings and start a garden that way.

Along with plants, there's another food resource that is surprisingly easy to get for free or cheap—chickens. We've discussed how chickens can be kept in a variety of situations. This sort of arrangement is becoming more commonplace and can be very easy to manage with just a small backyard. You can feed your chickens leftovers and water them with collected rainwater. When sourcing chickens, check the internet. Websites like Craigslist and Facebook can lead you to people giving away chickens or selling them for a low price so long as you are willing to come get them.

Tip 4: Reconsider your job.

For many people, a full-time 9–5 job is the pinnacle of financial stability. However, there are many times when a job like this can hold you back. The rigid hours can leave you locked into a daily schedule that leaves you no time to work toward the life you want. Some jobs can also be physically and mentally draining, leaving you with little energy or motivation. It's hard to focus on growing your own food when every day leaves you so tired that all you want to do is order fast food and be a couch potato.

Of course, it can be very difficult to leave a job, and I'm not advocating you just quit on a whim. Start by prioritizing your time; when you're on the clock, you're at work, and once you're punched out, that time belongs to *you*. Schedule time for your homesteading pursuits as if they were any other organized activity, and consider how you might start moving to a more flexible schedule. Could you go down to part-time for a while so you can build up a side business that you enjoy pursuing? Are there jobs available that you could do from home? Can you save for a year and have enough to support yourself while you transition to a more flexible job? Only you know the answers to those questions, and they will be slightly different for everyone. The point is, don't assume that a regular full-time job is your best option for a fulfilling and financially stable life.

Tip 5: If you can't own, rent.

It may seem as if you could only build on land that you own, but it is more common than you think for farming land to be rented. Usually, these rentals are on annual leases. You could potentially do a hybrid of renting and owning. If, for instance, you own a small amount of land with your house on it, but you're hoping to grow your operation, you could rent land from a neighbor for growing crops or grazing animals. If you're planning to do this, make sure your agreement is clear on who is responsible for what in regards to the maintenance of the property.

Another thing to consider is renting with the intention of buying. This is a specific kind of lease that you sign with a landlord, in which you agree to rent the property for a set amount of time, and when that time is up, you have the option to buy it. Not all landlords will agree to this, but if there's land that you're interested in, it is worth asking! Look for land that has been on the market for a while, as the owner might be more willing to consider a rent-to-own agreement. This is another time when imagination is a good resource, as properties that haven't sold aren't usually the most conventionally desirable ones. However, they often have a lot of promise if you're willing to put in some work.

There are a few things to be aware of in rent-to-own situations, and you will usually be required to pay a fee at the start of the lease. You should look over the terms carefully (as with all legal agreements) and make sure that the lease specifies things like who will be responsible for maintaining the property and how the eventual purchase price will be determined. Sometimes, the monthly rent you pay will go toward the price of the property, but sometimes it will not—this should also be clear in the lease. Be aware of the terms "lease-option" and "lease-purchase." Lease-option means you have the option to buy at the end of your lease; lease-purchase can mean that you are definitely agreeing to purchase. Again, read carefully, and if you're not sure about the meaning of anything, consult a lawyer!

Tip 6: Get a loan.

As with any loan, proceed with caution and only if you have a plan in place to be able to repay it. There are three specific kinds of loans for buying land without property on it. Raw land loans are for land with absolutely no development, improved land loans are for land with standard developmental markers such as road access and utilities, and unimproved land loans are for land that sits between those two, with some incomplete development. It can be very difficult to get any of these kinds of loans, and your best bet is probably small, local banks or credit unions that you have an existing relationship with. You'll also likely need great credit and proof of savings or other collateral.

Next!

One of the biggest expenses in your budget is usually housing, and this is especially true when you're looking at buying a new house or property. While you can homestead from anywhere and in any type of housing, if you're planning to buy a new place, you might have some questions and concerns. The next chapter will help answer and alleviate those by giving you an overview of homesteading properties.

8

DIFFERENT TYPES OF HOMESTEAD PROPERTIES

Buying a new home can be very exciting, but it can also be incredibly stressful and frustrating. One of the best ways to alleviate the stress of this process is to be very prepared before you even start looking at properties. If you've made yourself a budget using the tips in the last chapter, then you hopefully have a good idea of what you can afford. The next step is to get a clear idea of what your ideal property looks like. While it's unlikely that you'll be able to find somewhere that exactly matches your ideal, knowing what you want helps you eliminate what you don't want. I recommend making two lists, one of the qualities that are absolutely necessary—things you cannot live without—and one of the things you would like to have if you could. Start with the first list, and use it to narrow down your options. Then, if you're deciding between properties, you can use the second to help you pick the best one for you.

Timing is also important when buying a new property. The optimal time to shop for a home is when you're financially able to buy, but not in a rush. You don't want to feel that you *have* to buy, as that might lead you to buy somewhere that doesn't really fit your needs. Remember, this will be your home, so it will affect all other parts of your life. At the same time, don't get hung up looking for the "perfect" property. It doesn't exist! At least not yet—that's where you come in. If you can find somewhere that meets all your basic needs, then you can put your skills to work turning it into the perfect place for you. This chapter will help you know what to look for in a property, what types of properties are popular among homesteaders, and give you tips for where to find them and how to go about buying them.

Types of Housing

I want to introduce you to some housing options that you might not be familiar with, but remember that many homesteaders do live in traditional houses. If that's what will work best for you, that's what you should do. But if you're interested in what other options exist, this list will give you an overview of some of those possibilities.

Cabins

Many people are familiar with cabins as recreational areas or second homes, but they can also be used as primary residences. Cabins are a good option if you're hoping to build with local materials, especially if the land you're building on is wooded. Some cabins are very basic, while others are designed to offer all your modern conveniences in a smaller and more rustic unit. If you're interested in full-time cabin living, one tip is to look for a cabin rental that has a comparable level of off-grid living to what you desire. By renting the cabin for a few weeks, you can determine if that lifestyle works for you. You can even do this a few times, trying out different lengths of stay or in different seasons, to really get a feel for the way of life before fully committing to it.

Tiny Houses

Tiny houses have been an increasing trend for a few years now. You might have seen them on Instagram or HGTV. A tiny house compresses all the basic needs of a house down into a much smaller package—usually between 100 and 400 square feet. While this might seem impossibly small (especially to Americans, who are used to square footage about six times as large), downsizing to a tiny house helps many people discover that they really didn't need as much space as they thought. If you have a large family, a tiny house might not be the best idea, but if you're living alone or as a couple, they can be a great option—especially if you don't have a lot of land, or if you're planning to share the land with other residents. The small size also means less money for building materials, and in some places, the size can mean a tiny house is not restricted by normal building codes. Some tiny houses are designed to be easily mobile, meaning that you could travel with them, or they could go with you to a new property. This can be a great option if you don't need much space now but might like to scale up in the future; your tiny house could be moved to a larger property and used as a living space while your next home is built.

Ecocapsules

Ecocapsules are a specific brand of tiny house or "micro-home" designed by Nice Architects, an architectural company in Slovakia. Ecocapsules come with all the benefits of a tiny house, but with additional considerations for environmentally friendly living. Ecocapsules are around 100 square feet and have built-in systems for wind power, solar power, and rainwater collection with filtration. They are very portable and can be placed on a trailer to be towed easily behind a vehicle. One of the biggest positives of buying an Ecocapsule is that you will not have to build it yourself. However, the unit costs around €80,000 (about $97,000 USD), with additional costs for shipping and transportation to your property. If you like the idea of an eco-friendly tiny house and don't want the hassle of building it yourself, this could be a great option. However, if you'd rather have a house that is customized to your specific needs and aesthetics, you might prefer to just design and build your own tiny house.

Earthships

Earthships are another eco-friendly type of housing. The concept of Earthships started with the vision of architect Michael Reynolds and the company that he helped found, Earthship Biotecture. However, many people use the term to refer to any type of housing built in that same style. This includes being made of natural and/or recycled material and operating at least partially off-grid. There tends to be more variety among Earthships than many of the other types of housing discussed on this list because Earthships are not about a specific structure or style, but about using local materials to create a home that is sustainably built and maintained. Earthships range from the modest and practical to the large and luxurious. If you're feeling particularly adventurous, you can design and build an Earthship on your own, but you can also get plans from or have a consultation with Earthship Biotecture if you'd prefer to have the more complicated parts of the process done for you.

Shipping Containers

If you'd love to build your own home but you want to cut down on the new materials you use, you might consider converting a shipping container. You can use one retired shipping container to create a small house for yourself, or you can join or stack several together to make a bigger living space. In addition to your basic structure already being made, shipping containers offer good shelter and can be transported easily. Their rectangular space mimics the layout of most houses, and you can design the space inside as an open floor plan or add room separators for a bit more of a traditional feel. One thing to be aware of with shipping container homes is that being made of metal makes it hard to regulate the temperature inside. In particular, if you live somewhere that has extreme cold or heat, you might want to try a different type of housing or put some research into ways to securely insulate the space. You'll be much more comfortable, and your heating system will thank you!

Underground Houses

On the flip side, a type of housing known for its incredible insulation is underground housing. The idea of underground homes might bring to mind images of Hobbit holes, but in reality, a lot of people find underground housing to be an efficient and practical way of life.

Because your living space is surrounded by earth, it is naturally insulated. Another benefit of underground living is the automatic level of privacy that you'll get—but don't worry, you can still have windows and plenty of natural light if you design your home properly. The level of design and planning for these homes is probably the biggest drawback, as they need to be able to bear the weight of the earth without collapsing inward. They also need to be built in a way that doesn't negatively impact the surrounding environment. This can be intimidating, but it also often leads to some innovative designs that fit beautifully with the natural world around them.

Yurts

This tent-like home appeals to people who want to have as little impact on the land as possible, given that it sits on a platform and is built from a wooden frame and lattice walls covered with fabric. This also makes it one of the more affordable options on this list. However, yurts are definitely not for everyone; they are more durable and luxurious than regular tents, but residents still contend with much of the same drawbacks as tent campers. This includes less insulation (from both weather and noise), less privacy, not much storage, and a lack of security against bugs and animals. Just like some people enjoy camping and some do not, for some people this sort of living is fun and interesting, and for others, it is completely unappealing. I trust you'll know which of these you are!

Where to Find Properties

No matter what type of housing you choose, you'll need somewhere for it to go. Whether you're looking for a fully developed farm or an undeveloped plot of land, you need to start your search somewhere. It can be good to seek the help of a real estate professional, especially if you're not familiar with the process or you're feeling overwhelmed. However, even if you are working with a realtor, I recommend also doing some searching on your own. You never know when you might find something your realtor missed, or come across an option that you hadn't considered previously. And of course, if you're not working with a realtor, knowing where to look for listings is even more important.

Mainstream Real Estate Sites

These sites, such as Zillow, Realtor.com, and Trulia, will have a wide range of properties, so the challenge with them is to narrow down your options. Most of these sites have a similar collection of filters that you can use for this. Use the home type filter to eliminate any unsuitable options such as apartments, condominiums, duplexes, and townhomes, and use the lot size filter to look for properties in your preferred size. Checking at least an acre will remove most urban and many suburban properties, and leave you with more rural options. The higher the minimum acreage you set, the more of your search results will be rural. Another thing you can do is look for an off-grid option among the filters—try looking for an option that says "more filters" or similar if you don't see it right away.

Sites That Specialize in Off-Grid, Sustainable, or Homesteading Properties

These sites have done the work of narrowing the field for you, and only offer properties in their specific niche. Some will cover many different aspects of homesteading and off-grid living, such as eco-friendly, sustainable, alternative energy, and survival properties, while some may narrow their focus to just one or two of these. Two sites you can start with are Sustainable Properties Real Estate Listings and Survival Realty. Sustainable Properties has property listings throughout the United States, and Survival Realty has listings in the U.S. and a small amount of overseas properties as well.

Social Media

I wouldn't recommend this as a first or only choice for finding real estate, but it can work to expand your options or if you've had no luck and you're running out of options. You can always follow local real estate offices and brokers on social media, as they will often post about new listings. If you're in a particularly competitive market, you can alter your settings to send you notifications whenever these accounts post, which can help you get the jump on any new properties. Additionally, sites that have groups or communities, such as Facebook or Reddit, can work. Search for off-grid or homestead real estate and see what groups come up. You can then bookmark or follow these groups so that you can check them regularly to see what postings have been added.

Questions to Ask When Buying Property

I encourage you to make your own tailored list of questions in addition to these, and of course, you'll also have the list of necessities that you've made. Be sure to review all of those whenever you look at a property. But the following list of questions will make sure you remember the basics and can help jump-start your thought process as you contemplate what you need from a property.

How much land do you need?

This is really a question you ask yourself, and the answer will be entirely dependent upon your household needs. You might have a firm minimum or maximum amount of land, or you might be flexible to a certain degree. The important thing is to know what size of lot is acceptable to you. To help you estimate how much you'll need based on what you plan to do, here is a quick rundown:

- 5 acres is enough for a garden whose crops will feed an average-sized family, plus space to raise small and medium livestock.
- 10 acres is enough for a garden to feed a family and

possibly sell the excess, as well as small and medium livestock plus a small herd of cattle or a few horses.
- 15 to 20 acres is enough for all of these above, plus growing hay and at least some of the food your livestock will need; if it is wooded, it might also supply most of your fuel needs.
- 25 to 50 acres is enough for a garden, whatever livestock you'd like, hayfields, wood for fuel (if in a wooded area), a barn, and possibly other structures (such as a greenhouse or a butcher shop), and growing crops. It will also likely have a much higher amount of natural resources.

Where is it located?

We've discussed the general location of homesteads in some depth already, and the importance of location in regards to laws and climate. Now that we've gone over all the different possibilities for your housing, your utilities, and many other choices you'll be making, you can see how choosing the right location affects all other areas of your life. Once you know the area where you're hoping to buy, make sure you look into the planning and zoning laws for that area. We've also discussed the benefit of a location with a lot of natural resources and the pros and cons of living in more remote locations versus more populated areas. Considering the location of a property also includes noting how far away you are from services that you might need, such as medical care. If the land is not developed, you'll want to inquire about the possibility of adding things such as a phone line or internet. You should also ask questions about the history of the area in terms of extreme weather and natural disasters.

How will you get there?

I don't mean just how you will get to the plot of land itself, but also to the place where the house sits or where you'd hope to build. If you're looking at relatively populous areas, the distance from the main road to your house might be short. But if you're looking at large amounts of land and/or in rural areas, the actual location of the house might be far away from publicly maintained roads. If there is a private road through the property, is it in good shape? What will it cost to maintain? Are you required to give anyone else access to it? Don't forget that in addition to getting yourself to and from the main road, you'll also need to transport supplies. This principle also applies to anywhere else that you'll need to get to on the property. If there is no legal, safe way to access the places on the property you need to go, there is no point in owning the property.

What's nearby?

This includes both residential and commercial properties. You should try to find out what is happening on adjacent plots of land and whether any of those activities will affect *your* land. Pay particular attention to any farms that are uphill or upwind from the property, as anything that enters their air or water can make its way to your air and water—especially if there is a water source that runs through both properties. Keep in mind that undeveloped land near yours might be developed in the future. You can check to see what zoning and restrictions exist for the area as a whole in order to get an idea of what it could or couldn't be used for in the future.

You can also ask "Who is nearby?" If possible, it is always good to see if you can meet the neighbors and get to know them a little. If you're lucky, you can find a place that you like where you also like the neighbors.

What restrictions are there on usage or rights?

As I briefly mentioned when discussing water systems, sometimes you can own land but not have the rights to the water on that land. There can be similar issues around mineral rights, and some deeds can also include restrictions on how the land can be used. Whether or not these things are deal-breakers is up to you, but it's important that you ask the question so you're making an informed decision. One thing to specifically look out for is land with covenants. Covenants are sets of rules that are designed to keep the usage of an area very uniform, and will often prohibit common homesteading practices such as gardens or livestock. Some might also dictate what types of structures you can build. As you consider all these restrictions, also consider what you might want to do with your property in the future. As you become more adept, you might want to expand your operation. Just because you don't want to raise livestock now doesn't mean you won't in the future. Think about long-term possibilities, and don't lock yourself into a limited range of activities just because you don't want to do any of those things immediately.

How well is the house insulated?

We covered the importance of insulation when discussing energy efficiency. The better the insulation, the easier it will be to heat and cool your house. If the house you're considering doesn't have proper insulation, you'll need to consider if improving it is something you'd be willing to do. You should be sure that what you end up paying for a property reflects the extra cost you'll incur for those improvements.

What is the land itself like?

Look at the way the land is laid out, especially in the context of how you plan to use it. For example, if wind power plays a major role in your plans, is there somewhere suitable to put your turbine? If you're planning to have fields for crops or for grazing, how difficult will it be to establish those? If there's a water resource you want to use, how far is it from the house (or where you expect to build)? If you're planning to hunt or forage for food, are those resources available on the land? Customize these types of questions to your personal plans for the property. Remember that you can change many things about a property once you purchase it, but those changes will take plenty of time and money.

What is the soil like?

Because almost all forms of homesteading involve growing food, it's a given that the quality of the soil will affect you directly. Avoid areas with a lot of clay or sand in the soil. As we discussed in chapter four, the ideal type of soil is rich and not too hard or compact, with good drainage and a balanced pH.

You can use the tips from that chapter to examine the soil and determine how good it will be for growing. You can also check the USDA.gov website for more information about the area. Their site has soil maps and data for much of the United States.

Where does the water come from?

Whether you're buying undeveloped land or a full compound, there is a version of this question you need to ask. If there is already a water system in place on the property, you need to know what it is and how it works. If there is not, what water resources exist? Are they suitable for the lifestyle you're planning? You need to know if all this a) fits your needs, and b) needs any alterations or repairs for it to work as you'd want. Also remember to have the water tested by a state-certified lab before you buy, as we covered in the well water section of chapter three.

Have you done a title search?

If you've found a property you really like and you're strongly considering buying it, make sure that you look into the public records for the property. You can pay to have this done, or you can do it yourself. You want to be sure there are no encumbrances or liens on the property because those will become your problem once you are the owner. (An encumbrance is anything that dictates how a property can be used, including whether or not it can be sold; a lien is a specific type of encumbrance where a creditor lays claim to part or all of your property because of money owed to them.) It is also a good idea to just be sure that all the details of the property, including the owner's identity, are correct as they've been given to you. You might think that this is overkill, and it's true that most of the time these searches will turn up nothing. However, there are people who will be dishonest and try to hide these things, and you don't want to find out that's the case after you've bought the land. This is a simple step you can take that can save you a lot of time and hardship.

Next!

I've given you a lot of information to process, but I've also tried to contextualize that information, giving you the why and how in addition to the what. We have one last thing to talk about, and it's perhaps the most important context of all—the real-life context of a homesteading community. None of the decisions you make will happen in a vacuum, and the next chapter will discuss not only the importance of your community but how to get the most from it.

9

BEING PART OF A HOMESTEADING COMMUNITY

If you're particularly observant, you might have noticed that the idea of community has already come up many times in this book. We talked about it when we debunked the common myth that homesteading is about living in the middle of nowhere, surrounded by wilderness and with no company but your own thoughts. There are, of course, those people who will seek out extreme isolation—but for most, neighbors are a reality. And hopefully, they are not just a reality, but a welcome benefit.

This is why the idea of neighbors came up when we discussed plants and livestock, again when we talked about emergency preparation, and again when we went over what to look for in a property. The principle of returning to a more connected way of life that is central to most people's homesteading journey is not just about connecting to nature but also connecting to people. More conventional life is often about depending upon systems to help you in times of need—systems that can and do fail people. This life is about depending on yourself, yes, but it is also about developing a network of like-minded people that you know and respect. It is rebuilding society to its ideal of working together to make sure everyone can survive and thrive. Here are some reasons that you may want to find a local homesteading community.

The Benefits of Having a Homesteading Community

Safety

The old adage that there's strength in numbers exists for a reason. Safety is a big part of why societies evolved in the first place. Our ancestors all got together and decided that it was easier to fight off predators and survive catastrophes if they were pooling their resources. The same is true of modern communities, and especially true in homesteading. Having a good relationship with the neighborhood will mean that you can all look out for each other. You can report suspicious behavior, and watch each other's homes when they're empty. Even something as simple as a neighbor noticing a light on when you said you'd be gone and calling to check that everything is okay can avert crime and keep everyone safer.

Cups of Sugar (and Other Supplies)

This is another well-known phrase associated with neighbors: "Can I borrow a cup of sugar?" Even if you do your best to prepare and plan, there will be moments when you discover you need something that you don't have. It might be what you need to make that cake you were planning, or it might be a tool that breaks when you have an important task to do. If you have the kind of rapport where you can walk or call over to the person who lives next to you and ask them for help, they might be able to save the day. And then, of course, you can return the favor next time they need something. This can extend beyond supplies to include actions, as well. Maybe you have a package to mail but you can't make it to town, but you know your neighbor goes in every week for a standing appointment and would be willing to mail your package when they go. In return, you might pick up or drop off something for them the next time you run errands.

Advice

We've already discussed how when you first move to a new location, your neighbors will be some of the most useful sources of information for you. They will be familiar with the area and have tons of accumulated knowledge that they will likely share if you just ask. Even after you have been homesteading for years, you will find that some neighbors have areas of expertise that you can draw on. You yourself will probably develop specialized skills that will be useful to others. It is impossible for one person or household to be good at everything or know all there is to know, but with every household sharing its own particular niche, a community can fill in gaps and ensure that everyone has what they need.

Barn Raising (and Other Communal Efforts)

It was common among communities in the past to hold a barn raising whenever someone needed a new barn. A barn raising was an event where everyone in the community came together to build a barn (or some other structure) for someone, with the understanding that the community would do the same for them when it was their turn. This idea doesn't just apply to barns! There are so many aspects of homesteading that will go much faster if done with others. Combining forces to make bigger and better things happen is a time-honored tradition, and it doesn't even have to be task-based. Perhaps there's an expensive piece of machinery that you'll only need for a small period each year? Ask your neighbors if they want to purchase it together—sharing the machine means sharing the cost, which is a win for everyone.

Entertainment and Socialization

Community is not just a practical consideration. It also fulfills the very human part of us that wants companionship and enjoyment. Having get-togethers with your neighbors is a great way to get to know one another, and also a great way to have some fun. Particularly if you are raising a family, the chance to meet up with other families and let the kids play while the adults chat can be a great way to blow off steam and relieve stress, whether it's a large, organized event or just a small gathering.

Social Networking

In the era of the earliest homesteaders, families often lived very close to one another, and this extended network was often utilized to help out with everything from childcare to getting your mail while you were out of town. Nowadays, it is much more common for people to live very far from their families, in different neighborhoods, towns, states, or even countries. However, that doesn't mean that you can't find a network of people willing to look out for one another in those same ways. Many people find that after years of building relationships with their neighbors, those people have become a kind of family to them. Perhaps even more beneficial than the favors themselves, this sense of found family can be a boon to our mental and emotional well-being.

Commercial Networking

However you plan to make money on your homestead, it's a given that you will need contacts to do it. One of the best ways to make those contacts is to reach out to the people around you. Everyone in your circle will also have their own circle of people who might want the products or services you're offering. If people know you and like you, they are more likely to support your business. Neighbors who have been in the area for a long time might know just the right person to put you in touch with, while even neighbors who have nothing to do with your business can help you by promoting what you do to others. Again, this sort of thing is dependent upon establishing and maintaining relationships. You can't get just anyone to do these things for you—it will be the people who feel a positive connection with you that will go the extra mile to help you out.

Help in Times of Need

This is a thread that runs through most of the other items on this list. At its best, a community is a system that you can rely on to help you when you need it. This is because by building this community, you have essentially enacted a social contract or an unspoken agreement that you will all work together to make life easier and better for one another. At no point is this more crucial than when something goes wrong. If you get an unexpected illness or injury, it could seriously hamper your ability to work for days, weeks, or even months. However, while you're recovering, work still needs to get done. Having a good neighbor or several who are willing to drop by and pick up even small tasks, such as feeding livestock or watering your garden, can make all the difference in times like this. Just as I've pointed out several times throughout this book, small changes in your life can accumulate into big differences; this rings true here, as small acts of kindness and aid can add up to make the difference between success and failure. This is the biggest benefit that you can get from your community.

Principles for Peaceful Living

Now that we've established how valuable your community can be, let me give you a few tips for making that community as strong as possible. There are four principles that I always do my best to live by when interacting with others, particularly those in my own community. Putting these into practice will help you create the kind of harmonious coexistence that you and your neighbors will appreciate.

Communication

If you only pay attention to one of these principles, make it this one. A huge amount of conflict boils down to miscommunication—don't let that happen to you! The two pillars of communication are clarity and frequency. Clarity means making sure that you are expressing yourself clearly and that your neighbors always know what to expect from you and what you expect from them. Explaining why you did something or want to do something can mean the difference between your neighbor resenting you or them helping you out. Frequency means maintaining the lines of communication through consistent contact. Don't just talk to your neighbor when there's a problem or when you need something; try to build a valuable relationship with them, as well.

Respect

This goes hand in hand with communication since one of the best ways to convey that you respect someone is to give them the courtesy of a warning about any decisions you are making that might affect them. This might be as small as a heads up that you will be making noise at a certain time during the day, or as big as informing them that you will be putting up a wind turbine that they will be able to see from their property. (Bonus points if you give them the consideration of asking if they will be okay with big changes that affect them. It's your property, you don't have to—but doing so can go a long way toward making the relationship work.) And always respect the boundaries of your properties! Even if they've allowed you to cross or use their land in some way in the past, *always* be sure to ask before doing it so they know you respect their ownership of the land.

Compromise

This also goes hand in hand with communication. (Are you sensing a pattern? Communication is key!) Sometimes in a tense situation, you can figure out a solution where everyone gets exactly what they want. But much more frequently, conflict arises because there is no one solution that is perfect for everyone. In these cases, acknowledge that there is no bad guy, but rather that everyone involved is trying to do what they feel is best for them. To find the best compromise, carefully consider your own priorities. Draw the line between what you need to happen and what you would like to happen, and then clearly communicate those: "X is what I absolutely need to happen, but I'm flexible about Y." You'd be surprised how often offering to give something up will prompt the other person to do the same.

Generosity

Just as offering to compromise can engender goodwill in a conflict, offering something of value to your neighbors can foster a sense of community. There are many different ways you can put this into practice. It might be goods such as eggs, vegetables, textiles, or meat; it might be time and energy, such as helping them to put up a fence or with shearing their sheep; or it might even be space, such as sharing some of your garden space or a field you aren't using. Offering something of value without expecting anything in return illustrates that you are interested in the good of the community as a whole, not just your own self-interests. And the truth is, the wellbeing of the community *is* in your self-interest. The stronger the community, the better it will be able to take care of its own.

CONCLUSION

At this point, I've covered all the basics you'll need to get you started on your homesteading plans. Let's take a moment to think back through everything you've learned.

First, we debunked some common myths and discussed all the different ways you could incorporate homesteading into your life. After this, we went over common legal issues and I walked you through the best states in the United States for homesteading. Next, we got into some of the nitty-gritty details of different aspects of homesteading: the options you have for water, power, waste, and heating; how to start gardening and raising livestock; how to be prepared for medical emergencies; how to keep your homestead in good working order; and how to create and stick to a budget. And finally, I introduced you to different types of housing, gave you tips on how to find property of your own, and gave you some perspective on how fostering community is a crucial part of homesteading.

Things to Remember

Now that you've made it to the end of the book, I hope you'll take some time to digest everything you've learned. There's a lot of information here, and it will take some time to absorb all of it. While I hope everything I've included will be of use to you at some point, I want to remind you of the most crucial things to remember:

Planning and preparing can save you time, money, and frustration.

Any endeavor you're undertaking, from growing herbs in a pot on your balcony to running a 100-acre farm, can benefit from a clear plan. Do your research and make informed decisions. Not only are informed decisions easier to make, but they can save you from expensive and time-consuming mistakes.

Design your life for you, not anyone else.

Informed decisions aren't just about knowledge of homesteading; they're also about knowledge of yourself. Don't forget to start with your vision of your ideal life, and let that vision dictate what you prioritize. The most perfectly planned and executed life is meaningless if it's not the life that *you* want to be living.

Anyone can incorporate homesteading into their life.

A lot of the more granular information in this book, such as buying property and installing self-sufficient utilities, may seem like it is only applicable to large-scale homesteading. However, the principles of these things can be applied at any level. If you're not currently able to become as self-sufficient as you'd like, practicing the tenets of homesteading on a small scale will help you develop the skills and mindset for when you *are* able to make the change.

The best way to make lasting change is to take small steps over time.

You don't need to jump headfirst into a new lifestyle. In fact, you shouldn't! It is *much* more sustainable to change your life a little bit at a time. Use your SMART goals (look back at chapter one for a refresher) to slowly achieve what you want to achieve; it will be much less stressful and *far* more permanent.

Nothing will happen until you make it happen.

This is the most important thing to remember! Your life is in your hands, and the only one that can change it is you. I've given you the information, but it's up to you to act on it.

And that's it! Now go forth and start homesteading. I hope that this book has been of help to you. If it has, consider leaving us a review on Amazon—it'll help even more people find this book and start planning their own homesteading dreams!

NOW FOR SOMETHING SIGNIFICANT

Community is a massive part of the homesteading lifestyle, so please **help your fellow homesteaders to find this book by giving a one-click review and/or a couple of sentences.**

The information available in the vast and confusing online world is often overwhelming, missing points, or even incorrect. Your contribution by reviewing means that others in your position get the best information available.
You get to be part of a grassroots establishment aimed at helping each other live the best homesteading lifestyle possible.

Simply follow the link by scanning this QR code and leave your review

It warms my heart that you have read my book, and I look forward to seeing you in our online community.

A SPECIAL GIFT TO MY READERS

Included with your purchase of this book is your free copy of
Your Homestead Planner

Follow the link below to receive your free copy:
www.kellyreedauthor.com
Or by accessing the QR code:

You can also join our Facebook community
Homestead Living & Self Sufficiency,
or contact me directly via kelly@kellyreedauthor.com

REFERENCES

15 Acre Homestead. 2018. "Getting Started Homesteading: Budgeting." Accessed June 6, 2021. https://15acrehomestead.com/getting-started-homesteading-budgeting/.

Advanced Water Solutions. 2017. "How Does Water Distillation Work?" Accessed June 22nd, 2021. https://advancedwaterinc.com/water-distillation-work/.

American Dairy Goat Association. 2020. "ADGA Breed Standards." Accessed June 10, 2021. https://adga.org/breed-standards/.

The American Goat Federation. n.d. "Meat Goats." Accessed June 10, 2021. https://americangoatfederation.org/breeds-of-goats-2/meat-goats/.

REFERENCES

American Red Cross. n.d. "Make a First Aid Kit." Accessed June 5, 2021. https://www.redcross.org/get-help/how-to-prepare-for-emergencies/anatomy-of-a-first-aid-kit.html.

Ali, Shahraz. 2020. "Top 7 Survival Foods to Stock – When You Are Living Off-Grid." Off Grid Living. Accessed June 5, 2021. https://offgridliving.net/top-7-survival-foods-to-stock-when-you-are-living-off-grid/.

Atkins, Gordon. n.d. "Living off the Grid: Legal or Illegal?" The Homesteading Hippy. Accessed June 4, 2021. https://thehomesteadinghippy.com/living-off-the-grid-legality/.

Barnes, Steve. n.d. "6 Seldom Followed Tips for How to Buy Off Grid Land." The Off Grid Cabin. Accessed June 6, 2021. https://theoffgridcabin.com/the-6-must-know-tips-for-how-to-buy-off-grid-land/.

Bernard, Murrye. 2020. "Earth-Sheltered and Underground Homes Basics." The Spruce. Accessed June 15th, 2021. https://www.thespruce.com/what-are-underground-homes-1821786.

Brendza, Will. 2019. "How To Build A Root Cellar From The Ground Up For Survival." Skilled Survival. Accessed June 5, 2021. https://www.skilledsurvival.com/underground-food-storage-root-cellars/.

Brooke, Nick. n.d. "The Best Plants for Aquaponics." How to Aquaponic. Accessed June 22nd, 2021. https://www.howtoaquaponic.com/plants/best-plants-for-aquaponics/.

Brownlee, John. n.d. "Test Well Water Before Buying A Homestead – A Helpful Guide." Country Homestead Living. Accessed

June 22nd, 2021. https://www.countryhomesteadliving.com/should-well-water-be-tested/.

Burgess, Ross. n.d. "7 Common Misconceptions of Living Off the Grid." A Modern Homestead. Accessed June 2, 2021. https://www.amodernhomestead.com/misconceptions-living-off-the-grid/.

Cabin Life. n.d. "Composting vs. Incinerating Toilets." Accessed June 22nd, 2021. https://www.cabinlife.com/articles/composting-vs-incinerating-toilets.

Canadian Valley Electric Cooperative. n.d. "How Does Geothermal Energy Work?" Accessed June 22nd, 2021. https://www.mycvec.coop/how-does-geothermal-work.

Carlson, Riley E. n.d. "Beginner's Guide To Keeping Bees." Homesteading.com. Accessed June 5, 2021. https://homesteading.com/beginners-guide-keeping-bees/.

Carpenter, Dan. n.d. "Alternative Energy Solutions." Homestead Launch. Accessed June 5, 2021. https://homesteadlaunch.com/alternative-energy/.

College of Agriculture & Natural Resources. n.d. "Nucleus Colonies." University of Delaware. https://canr.udel.edu/maarec/nucleus-colonies/.

Copeland, Jayden. 2019. "What Is Homesteading And Is It For You?" Backroad Bloom. Accessed June 2, 2021. https://backroadbloom.com/what-is-homesteading-and-is-it-for-you/.

Copeland, Jayden. 2019. "11 Feasible First Year Homesteading Goals." Backroad Bloom. Accessed June 2, 2021. https://backroadbloom.com/2019-1-7-11-feasible-first-year-homesteading-goals/.

Counter, Angela. n.d. "Completely, 100 Percent Off-Grid: 9 Essential Foods You Should Grow." Off the Grid News Accessed June 5, 2021. https://www.offthegridnews.com/off-grid-foods/completely-100-percent-off-grid-9-essential-foods-you-should-grow/.

Crisis Times. n.d. "Raising Animals Off the Grid." Accessed June 5, 2021. http://crisistimes.com/offgrid_animals.php.

Culver, Blake. 2021. "How to Plan a Garden For 2021." An Off-Grid Life. Accessed June 5, 2021. https://www.anoffgridlife.com/how-to-plan-a-garden-for-2020/.

Davidson, Josh. 2021. "Off Grid Water Systems: 4 Proven Ways To Bring Water To Your Homestead." Tiny Living Life. Accessed June 5, 2021. https://tinylivinglife.com/2021/01/learn-how-to-build-off-grid-water-system/.

DeJohn, Suzanne. 2021. "Soak, Drip, or Spray: Which Is Right for You?" Gardener's Supply Company. Accessed June 22nd, 2021. https://www.gardeners.com/how-to/how-to-choose-a-watering-system/8747.html.

Department of Energy. n.d. "Solar Water Heaters." Energy Saver. Accessed June 5, 2021. https://www.energy.gov/energysaver/water-heating/solar-water-heaters.

Digital Public Library of America. n.d. "Primary Source Sets: The Homestead Acts." Accessed June 2, 2021. https://dp.la/primary-source-sets/the-homestead-acts.

Dodrill, Tara. n.d. "5 Off-Grid Water Sources and Systems." Homestead Survival Site. Accessed June 5, 2021. https://homesteadsurvivalsite.com/off-grid-water-sources.

Dodrill, Tara. n.d. "6 Features To Look For In Off-Grid Property." Homestead Survival Site. Accessed June 6, 2021. https://homesteadsurvivalsite.com/features-off-grid-property/.

Dodrill, Tara. n.d. "10 Best States for Homesteaders." Homestead Survival Site. Accessed June 4, 2021. https://homesteadsurvivalsite.com/top-10-states-living-off-grid/.

Drevets, Tricia. n.d. "8 Ways To Generate Power Off Grid." Homestead Survival Site. Accessed June 5, 2021. https://homesteadsurvivalsite.com/generate-power-off-grid/.

Earthship Biotecture. 2020. "Super Sustainable Buildings via Thermal Dynamics & Passive Solar." Accessed June 6, 2021. https://earthshipbiotecture.com/.

Ecocapsule. n.d. "Ecocapsule Home Page." Accessed June 15th, 2021. https://www.ecocapsule.sk/.

Ed., Old Farmer's Almanac. 2020. "Raising Chickens 101: When Chickens Stop Laying Eggs." The Old Farmer's Almanac. Accessed June 22nd, 2021. https://www.almanac.com/raising-chickens-101-when-chickens-stop-laying-eggs.

REFERENCES

Federal Emergency Management Agency. 2021. "Build a Kit." Ready.gov. Accessed June 11th, 2021. https://www.ready.gov/kit.

Ferguson, Donna. 2014. "Greywater Systems: Can They Really Reduce Your Bills?" The Guardian. Accessed June 22nd, 2021. https://www.theguardian.com/lifeandstyle/2014/jul/21/greywater-systems-can-they-really-reduce-your-bills.

Folger, Jean. 2021. "Rent-to-Own Homes: How the Process Works." Investopedia. Accessed June 14th, 2021.https://www.investopedia.com/updates/rent-to-own-homes/.

Gently Sustainable. n.d. "How to Homestead With No Money." Accessed June 6, 2021. https://www.gentlysustainable.com/how-to-homestead-with-no-money/.

Greene, Liz. n.d. "Trash on the Homestead: Everything You Need To Know." Homesteading.com. Accessed June 22nd, 2021. https://homesteading.com/trash-homestead/.

Happy Prepper. 2021. "Waste Disposal." Accessed June 5, 2021. https://www.happypreppers.com/waste-management.html.

Harbour, Sarita. 2021. "Off Grid Toilets: Which One Do You Want for Your Home?" An Off-Grid Life. Accessed June 5, 2021. https://www.anoffgridlife.com/off-grid-toilets/.

Harrington, Justine. 2018. "Regional Climates in the United States." USA Today. Accessed June 4, 2021. https://traveltips.usatoday.com/regional-climates-united-states-21675.html.

Harrington, Justine. 2018. "Climate Regions of the United States." USA Today. Accessed June 4, 2021. https://traveltips.usatoday.com/climate-regions-united-states-21570.html.

Haughey, Duncan. 2014. "A Brief History of SMART Goals." Project Smart. Accessed June 2, 2021. https://www.projectsmart.co.uk/brief-history-of-smart-goals.php.

Haynes, Sherry. 2021. "15 Best Goat Breeds for Meat." Pet Helpful. Accessed June 10, 2021. https://pethelpful.com/farm-pets/best-meat-goat-breeds.

Home, Garden and Homestead. 2021. "Soil Testing for Beginners." Accessed June 22nd, 2021. https://homegardenandhomestead.com/soil-testing/.

Hosfeld, Daniel. 2019. "Off Grid Electricity: What You Need to Know." An Off-Grid Life. Accessed June 5, 2021. https://www.anoffgridlife.com/off-grid-electricity-what-you-need-to-know/.

Howell, Elizabeth. n.d. "Five Kinds Of Off-Grid Living." HeroX. Accessed June 6, 2021.

Hunter, Jacob. 2021. "The Oh Crap! Guide to Off Grid Sewage and Septic Systems." Primal Survivor. Accessed June 5, 2021. https://www.primalsurvivor.net/off-grid-sewage/.

Incinolet.com. n.d. "Frequently Asked Questions." Accessed June 22nd, 2021. https://incinolet.com/frequently-asked-questions/.

Intrepid Outdoors. n.d. "How to Make a Bio-Filter." Accessed June 21st, 2021. https://intrepidoutdoors.com/make-bio-filter/.

Johnson, Jamie. 2020. "What's An Encumbrance In Real Estate?" Rocket Homes. Accessed June 22nd, 2021. https://www.rockethomes.com/blog/home-buying/encumbrance.

Johnson, Julie. n.d. "Everything You Need in Your Homestead First Aid Kit." Down to Earth Homesteaders. Accessed June 5, 2021. https://downtoearthhomesteaders.com/everything-you-need-in-your-homestead-first-aid-kit/.

Jones, Anna Newell. n.d. "12 Good (Financial) Reasons to Get to Know Your Neighbors." And Then We Saved. Accessed June 6, 2021. https://andthenwesaved.com/know-your-neighbors/.

Josephine, Val. 2017. "7 Ways to Power Your Homestead." Medium. Accessed June 5, 2021. https://medium.com/@Valsephine/7-ways-to-power-your-homestead-homestead-power-series-intro-a050bb0453be.

Just Dabbling Along. 2017. "10 Myths and Misconceptions about Homesteading." Accessed June 2, 2021. https://www.justdabblingalong.com/10-myths-misconceptions-homesteading/.

Kanuckel, Amber. 2021. "Companion Planting Guide." Farmers' Almanac. Accessed June 22nd, 2021. https://www.farmersalmanac.com/companion-planting-guide.

Lamp'l, Joe. 2018. "Raised Bed Gardening, Pt. 1: Getting Started." Joe Gardener. Accessed June 10, 2021. https://joegardener.com/podcast/raised-bed-gardening-pt-1/.

Lee, Shannon and Bob Vila. n.d. "5 Things Homebuyers Need to Know About Septic Systems." Bob Vila. Accessed June 22nd, 2021. https://www.bobvila.com/articles/septic-systems/.

Lewis, Patrice. 2013. "Homestead Water." Backwoods Home Magazine. Accessed June 5, 2021. https://www.backwoodshome.com/homestead-water/.

Loftsgordon, Amy. n.d. "What Is a Property Lien?" Nolo Legal Dictionary. Accessed June 22nd, 2021. https://www.nolo.com/legal-encyclopedia/what-property-lien.html.

Love Property. 2020. "Incredible Earthships: Off-grid Homes You've Got to See." Accessed June 15th, 2021. https://www.loveproperty.com/gallerylist/76795/incredible-earthships-offgrid-homes-youve-got-to-see.

Magyar, Cheryl. 2019. "45 Self-Reliant Skills Every Homesteader Needs To Know." Rural Sprout. Accessed June 2, 2021. https://www.ruralsprout.com/self-reliant-skills-for-homesteaders/.

Maxwell, Steve. 2012. "Homestead Water Sources and Options." Mother Earth News. Accessed June 5, 2021. https://www.motherearthnews.com/homesteading-and-livestock/self-reliance/homestead-water-sources-zm0z12aszkon.

McCafferty, Emily. 2017. "All About Off Grid Wastewater: Options, Septic, Code, and Advice." Accidental Hippies.

Accessed June 5, 2021. https://accidentalhippies.com/2017/07/25/off-grid-waste-septic/.

McCafferty, Emily. 2020. "What Does It Actually Mean to be "Off The Grid"?" Accidental Hippies. Accessed June 2, 2021. https://accidentalhippies.com/2020/02/17/what-is-living-off-the-grid/.

MelissaKNorris.com. n.d. "12 Tips on How to Raise Pigs for Meat." Accessed June 5, 2021. https://melissaknorris.com/howtoraisepigsformeat/.

Meyer, Sarah-Jane. 2019. "Your Home Maintenance Checklist." PrivateProperty.co.za. Accessed June 6, 2021. https://www.privateproperty.co.za/advice/lifestyle/articles/your-home-maintenance-checklist/6977.

Mitchell, Ryan. 2018. "How To Start Homesteading On A Budget." The Tiny Life. Accessed June 6, 2021. https://thetinylife.com/how-to-start-homesteading-on-a-budget/.

Morning Chores. n.d. "12 Things You Need to Know Before Getting Your First Ducks." Accessed June 5, 2021. https://morningchores.com/about-raising-ducks/.

Morning Chores. n.d. "Homestead Maintenance: Taking Care of Your Tools and Equipment." Accessed June 6, 2021. https://morningchores.com/homestead-maintenance/.

Morning Chores. n.d. "Homesteading with Neighbors: 6 Tips to Avoid Disastrous Conflicts." Accessed June 6, 2021. https://morningchores.com/homesteading-neighbors/.

Morning Chores. n.d. " Housing Your Chickens." Accessed June 22nd, 2021. https://morningchores.com/chicken-housing/.

MSPCA–Angell. n.d. "Farm Animal Health and Veterinary Care." Accessed June 21st, 2021. https://www.mspca.org/pet_resources/farm-animal-health-and-veterinary-care/.

National Geographic. 2012. "United States Regions." Accessed June 4, 2021. https://www.nationalgeographic.org/maps/united-states-regions/.

Nicholas, Nick. 2020. "Common Types Of UV Lamps For Chemical-Free Water Disinfection." Water Online. Accessed June 21st, 2021. https://www.wateronline.com/doc/common-types-of-uv-lamps-for-chemical-free-water-disinfection-0001.

Off Grid World. 2021. "Off Grid Living is Illegal! Sort of…" Accessed June 4, 2021. https://offgridworld.com/off-grid-living-is-illegal-sort-of/.

Off Grid World. 2021. "The Best Off Grid Heating Systems." Accessed June 5, 2021. https://offgridworld.com/the-best-off-grid-heating-systems/.

Off the Grid News. n.d. "The 3 Best Livestock For New Homesteaders." Accessed June 5, 2021. https://www.offthegridnews.com/how-to-2/the-3-best-livestock-for-new-homesteaders/.

Off the Grid News. n.d. "6 Quick Steps To A Debt-Free Homesteading Budget." Accessed June 6, 2021. https://www.offthegridnews.com/financial/6-quick-steps-to-a-debt-free-homestead-budget/.

Poindexter, Jennifer. n.d. 30 Best Cow Breeds for Meat and Milk You'll Want to Know About." Morning Chores. Accessed June 5, 2021. https://morningchores.com/cow-breeds/.

Poindexter, Jennifer. n.d. "Homesteading 101: What Is It and the Essential Steps to Get Started." Morning Chores. Accessed June 2, 2021. https://morningchores.com/homesteading/.

Rejba, Alex. n.d. "Living Off The Grid in the USA – Is It Illegal?" The Smart Survivalist. Accessed June 4, 2021. https://www.thesmartsurvivalist.com/living-off-the-grid-in-the-usa-is-it-illegal/.

Rhoades, Heather. 2021. "Plant Spacing Guide." Gardening Know How. Accessed on June 21st, 2021. https://www.gardeningknowhow.com/edible/vegetables/vgen/plant-spacing-chart.htm.

Robinson, Ed. n.d. "Raising Pigs for Meat." Mother Earth News. Accessed June 5, 2021. https://www.motherearthnews.com/homesteading-and-livestock/raising-pigs-meat-zmaz70mazglo.

Rural Living Today. n.d. "Best States for Homesteading: Know Your Options." Accessed June 4, 2021. https://rurallivingtoday.com/homesteading-today/best-states-for-homesteading/.

Schipani, Sam. 2019. "How to Prepare for Emergencies When You Live Off the Grid." Hello Homestead. Accessed June 5, 2021. https://hellohomestead.com/how-to-prepare-for-emergencies-when-you-live-off-the-grid/.

Schwartz, Daniel Mark. n.d. "Off Grid Water Purification: Safe and Low Cost." Off Grid Permaculture. Accessed June 5, 2021. https://offgridpermaculture.com/Water_Systems/Off_Grid_Water_Purification__Safe_and_Low_Cost.html.

Seaman, Greg. n.d. "Choosing Land for Homestead Living." Earth Easy. Accessed June 6, 2021. https://learn.eartheasy.com/articles/choosing-land-for-homestead-living/.

Sher, Savannah. n.d. "The Best Places in America for Off-Grid Living." Bob Vila. Accessed June 4, 2021. https://www.bobvila.com/slideshow/the-best-places-in-america-for-off-grid-living-578748.

Silbajoris, Alex. 2018. "What Are the Differences Between Bleach and Chlorine?" Sciencing. Accessed June 8th, 2021. https://sciencing.com/difference-between-bleach-chlorine-6516255.html.

Skilled Survival. 2019. "Best Non Perishable Food To Thrive During Times Of Turmoil." Accessed June 5, 2021. https://www.skilledsurvival.com/non-perishable-foods/.

Skilled Survival. 2019. "How To Build A DIY Aquaponics System For Food Self Sufficiency." Accessed June 5, 2021. https://www.skilledsurvival.com/diy-aquaponics/.

Tactical.com. 2020. "Homing In on Off-Grid Homes." Accessed June 6, 2021. https://www.tactical.com/offgrid-homes/.

Tamara, Nadia. n.d. "The Top 12 U.S. States for Homesteading." Crisis Equipped. Accessed June 4, 2021. https://crisisequipped.com/best-states-for-homesteading/.

Tomisch, Emma. 2021. "Land Loans: Everything You Need To Know." Rocket Mortgage. Access June 14th, 2021. https://www.rocketmortgage.com/learn/land-loans.

Unbound Solar. 2020. "What States Allow You to Live Off the Grid." Accessed June 4, 2021. https://unboundsolar.com/blog/off-grid-legal-states.

Vuković, Diane. 2021. "Best States for Homesteading". Primal Survivor. Accessed June 4, 2021. https://www.primalsurvivor.net/best-states-homesteading/.

Weather Atlas. n.d. "Monthly Weather Forecast and Climate Alaska, USA." Accessed June 4, 2021. https://www.weather-us.com/en/alaska-usa-climate.

Weather Atlas. n.d. "Monthly Weather Forecast and Climate Hawaii, USA." Accessed June 4, 2021. https://www.weather-us.com/en/hawaii-usa-climate.

Whittington, Amanda. n.d. "How To Keep Your Home Warm When Living Off Grid." Homestead Survival Site. Accessed June 5, 2021. https://homesteadsurvivalsite.com/keep-home-warm-living-off-grid/.

Winger, Jill. 2019. "Become a Beekeeper: 8 Steps to Getting Started with Honeybees." The Prairie Homestead. Accessed June 5, 2021. https://www.theprairiehomestead.com/2014/05/get-started-honeybees.html.

World Population Review. 2021. "Best States To Homestead 2021." Accessed June 4, 2021. https://worldpopulationreview.com/state-rankings/best-states-to-homestead.

Ygrene. 2020. "What Are Solar Water Heating Systems?" Accessed June 5, 2021. https://ygrene.com/blog/renewable-energy/what-are-solar-water-heating-systems.

Young, Olivia. 2018. "See inside the 104-square-foot tiny house that helped a Canadian photographer 'house-hack' his way into living for free." Business Insider. Accessed June 15th, 2021. https://www.businessinsider.com/tiny-house-minimalism-zero-waste-saving-money-2018-12.

Zaheer, Kinza. 2020. "Growing Your Survival Garden for Off Grid Living." Off-Grid Living. Accessed June 5, 2021. https://offgridliving.net/growing-your-survival-garden-for-off-grid-living/.

www.ingramcontent.com/pod-product-compliance
Lightning Source LLC
Chambersburg PA
CBHW070256010526
44107CB00056B/2478